T3-BSE-268

Market Demand for Dairy Products

Market Demand for Dairy Products

EDITED BY

S. R. Johnson, D. Peter Stonehouse, and Zuhair A. Hassan

Iowa State University Press / Ames

This book was made possible in part by the generous financial support of the John R. Jackson Foundation in Ottawa, Ontario, Canada. The late John Jackson was a long-time member and president of the National Dairy Council, a body representing the interests of dairy processing plants across Canada.

Iowa State University Press gratefully acknowledges the editorial management of this project by the staff of the Center for Agricultural and Rural Development, Iowa State University.

∞ Printed on acid-free paper in the United States of America

First edition, 1992

Library of Congress Cataloging-in-Publication Data

Market demand for dairy products / edited by S. R. Johnson, D. Peter Stonehouse, and Zuhair A. Hassan.—1st ed.
p. cm.
Includes index.
ISBN 0-8138-0289-X
1. Dairy products industry—Mathematical models—Congresses. 2. Demand (Economic theory)—Mathematical models—Congresses. I. Johnson, Stanley R. II. Stonehouse, D. Peter (David Peter) III. Hassan, Zuhair A.
HD9275.A2M37 1992
381'.417'015118—dc20 91-26898

CONTENTS

CONTRIBUTORS

DONALD E. AULT is an Economist at Land O'Lakes, Inc., Minneapolis, Minnesota.

EVELYN BELLEZA is a Graduate Research Assistant in Agricultural Economics, Department of Agricultural Economics and Rural Sociology, Auburn University, Auburn, Alabama.

JAMES BLAYLOCK is a Senior Economist, Economic Research Service, U.S. Department of Agriculture, Washington, D.C.

RUEBEN C. BUSE is a Professor of Agricultural Economics, Department of Agricultural Economics, University of Wisconsin, Madison, Wisconsin.

HUI-SHUNG CHANG is an Assistant Professor of Agricultural Economics, Department of Agricultural Economics and Rural Sociology, Auburn University, Auburn, Alabama.

MERRITT E. CLUFF is Chief of the Forecasting Unit, Market Outlook and Analysis Division, Agriculture Canada, Ottawa, Ontario, Canada.

THOMAS L. COX is a Professor of Agricultural Economics, Department of Agricultural Economics, University of Wisconsin, Madison, Wisconsin.

ELLEN W. GODDARD is an Associate Professor, Department of Agricultural Economics and Business, University of Guelph, Guelph, Ontario, Canada.

RICHARD D. GREEN is a Professor of Agricultural Economics, Department of Agricultural Economics, University of California, Davis, California.

GARRY R. GRIFFITH is a Senior Research Scientist, New South Wales, Department of Agriculture, Anmidale, New South Wales, Australia.

RICHARD C. HAIDACHER is the Leader of the Food Demand Research Section, Economic Research Service, United States Department of Agriculture, Washington, D.C.

M. C. HALLBERG is a Professor of Agricultural Economics, Department of Agricultural Economics and Sociology, Pennsylvania State University, University Park, Pennsylvania.

ZUHAIR A. HASSAN is Associate Director, Policy Branch, Agriculture Canada, Ottawa, Ontario, Canada.

S. R. JOHNSON is a Professor and Administrator, Center for Agricultural and Rural Development, Iowa State University, Ames, Iowa.

HENRY W. KINNUCAN is a Professor of Agricultural Economics, Department of Agricultural Economics and Rural Sociology, Auburn University, Auburn, Alabama.

RALPH G. LATTIMORE is a Professor of Agricultural Economics, Department of Agricultural Economics and Marketing, Lincoln College, Canterbury, New Zealand.

A. J. OSKAM is an Associate Professor, Department of Agricultural Economics, Wageningen Agricultural University, Wageningen, The Netherlands.

JOHN C. ROBERTSON is a Research Economist, Department of Agricultural Economics, Virginia Polytechnic Institute and State University, Blacksburg, Virginia.

RONALD A. SCHRIMPER is the Associate Head of Agricultural Economics Research and Teaching, North Carolina State University, Raleigh, North Carolina.

D. PETER STONEHOUSE is an Associate Professor, Department of Agricultural Economics and Business, University of Guelph, Guelph, Ontario, Canada.

TESFAYE TEKLU is a Research Fellow, Food Consumption and Nutrition Division, International Food Policy Research Institute, Washington, D.C.

APELU TIELU is a Graduate Research Assistant in Agricultural Economics, Department of Agricultural Economics and Business, University of Guelph, Guelph, Ontario, Canada.

R. E. WILLIAMS is the Director of Economics, Milk Marketing Board of England and Wales, Thames Ditton, Surrey, England.

PREFACE

IN 1985, a workshop organized under the auspices of the International Dairy Federation (IDF) was held in Ottawa, Canada. The theme of this workshop dealt with applications of and experiences with econometrics and other modeling techniques for demand analysis, forecasting, and policy evaluation in the dairy industries of IDF member countries.

In the process of publishing the proceedings from this workshop, the idea of also producing a book was proffered by one of the workshop participants. The appeal in this idea was in reaching a wider audience than that served by the IDF proceedings.

The purpose of the book is to provide an up-to-date information base on recent and prospective developments in the theory and empiricism of consumer demand. Although the theme dwells on market demand for dairy products, the discussions of theoretical underpinnings and problems of empirical techniques are equally applicable to market demand for other food products, and indeed to consumer demand in general. Because of their inherent characteristics, dairy products fortuitously provided an excellent illustrative example base from which to proceed.

The intended audience for the book may, therefore, be thought of as multifaceted. On the one hand are those academicians, researchers, students, and others interested in the latest developments in consumer demand theory and its empirical application through econometric models and other techniques. On the other hand are those involved in leading, counseling, developing, and otherwise nurturing dairy industries throughout the world. The latter group is meant to include industry participants and advisors in both the developed and developing countries. For those in developing countries, the book should provide appropriate guidance for promoting domestic dairy industries based on the experiences obtained in developed economies with more mature dairy sectors.

We should like to acknowledge others who have contributed significantly to the production of this book. One is the National Dairy Council of Canada, which provided generous financial support for the preparation and publication of the book. The other is Ms. Judith Gildner, who exceeded her technical editing duties by providing a constant source of encouragement and moral support throughout the undertaking.

S. R. Johnson
D. Peter Stonehouse
Zuhair A. Hassan
July 1990

Market Demand for Dairy Products

CHAPTER 1

Introduction

S. R. Johnson, D. Peter Stonehouse, and Zuhair A. Hassan

ACCURATE INFORMATION on consumption responses to changes in relative prices, income, and other scaling and translating factors is being increasingly recognized as critical to the dairy industry. From the viewpoint of producers and processors, these demand parameters are important for investment and production capacity planning, as well as in shorter term management decisions on stock levels, allocation, sales, and advertising and promotion expenditures. These same parameters play a key role in the design and evaluation of public policy for the dairy industry. In many nations, the dairy industry is regulated to maintain stable producer prices and incomes and the quality of the consumer product.

This book contains a collection of research results on the development and use of demand parameters for dairy products. A number of the chapters are based on presentations made at an International Dairy Federation Workshop on "Demand Analysis and Policy Evaluation" held in Ottawa, Ontario, during 1985. Based on the results of this workshop and a continuing interest in developing improved information on dairy demand parameters, a plan was developed for the book. A grant from the John R. Jackson Foundation in Ottawa, Ontario, Canada, made the effort possible.

The book provides the dairy industry and demand analysts with current results on demand theory, estimation methods for demand systems, data and other information available for use in estimating demand parameters, and, finally, examples of applications that utilize demand parameters in public policy and farm and market management contexts. In each area, innovative developments and new results are summarized, with special emphasis on carrying the associated concepts to issues of importance for the dairy industry.

Policy Analysis and Regulation

In many nations the dairy industry is regulated by government. In some cases, these regulations have been derived from past concerns with maintenance of quality for dairy products. More recently, the regulations have been rationalized on the basis of maintenance of producer income, consistent supplies, price stability, and insulation from variations in international markets. The regulatory context for the dairy industry is, however, in the process of continual change. Among the factors responsible for this changing regulatory structure are an increasing internationalization of the markets for dairy products, developments in production and consumption technology, and, perhaps as important, changes in the representation of dairy producers and processors in the political economy of national and international economic policy formation.

The changing regulatory context for the dairy industry and the role of demand parameters in the regulatory process are described in Chapters 2, 5, 6, and 15. Chapter 2 by Griffin, Lattimore, and Robertson describes the regulation of the dairy sector and the use of demand parameters in policy analysis from a multinational perspective. Themes identified are the deregulation of the dairy industry and the concentration of policies more on stability and supply management than on the isolation of domestic markets and income transfers to producers. Also emphasized is the changing technology for dairy production, processing, and consumption. Modern production technologies have led to increased farm size and fewer producers, decreasing the political representation of the industry. Transportation, advertising, packaging, and other processing and distribution technologies have led to national distribution systems and a concentration in this area of industry as well. Finally, the demand for dairy products has changed in response to health concerns, the age composition of the population, the changed food supply, and other factors. From a multinational perspective, the picture is one of policy structure for the dairy industry in change. Demand parameters are playing a key role in positioning dairy policies to achieve improved performance in the national sectors and in the international markets for dairy products.

The importance of demand parameters in policy analysis is illustrated further by results of a case study for Canada. In Chapter 5, Cluff and Stonehouse show how demand parameters are employed in the supply-management policy for the Canadian dairy industry. In this management scheme, annual and seasonal dairy demand are characterized by own- and cross-price elasticities and income elasticities of demand. These elasticities and the corresponding aggregate demand functions appropriately calibrated serve as a basis for determining estimates of market requirements at administratively set prices. Prices are in turn adjusted to reflect the cost of production. Accurate estimates of the demand parameters are shown to be important in regulating the Canadian dairy industry to achieve balances between supply and demand and to limit the variability of government outlays and producer assessments. Chapters 6 and 15 indicate the importance of dairy demand parameters in different policy and regulatory concepts. Advertising and product promotion have become increasingly utilized in the dairy industry. In the United States, for example, a producer checkoff program has generated significant revenue for generic advertising and market development. As well, there is significant brand-related advertising by the processors of dairy products. The policy issues raised by producer supported promotion of dairy products and, in part, advertising by the private sector are highlighted by Chang, Green, and Blaylock. In the United States and in other nations, the emergence of producer supported market promotion and its relationship to private advertising will pose important policy issues on the incidence of potential benefits, allocation of resources by producer representative groups, targeted segments of consumers, and so on.

Hallberg concentrates on the intermediate demand for dairy products. In the dairy industry, and generally in the agricultural sector, farm-level outputs are appropriately characterized as inputs to production processes that in turn yield outputs for the consumer market. The conceptual context for deriving the associated demand relationships involves a system of parameters at the consumer level, processing technologies and costs, and impacts of regulations from public policy for the dairy industry. Typically, the fluid milk not directly consumed is an input in the production of cheese, butter, ice cream, and other dairy products. These products have their own retail price and income elasticities. As well, cross-price effects are important in influencing their market demands. To complicate matters further, the manufacturing milk price is often derived from a regulated price for fluid milk. Estimates of the demand for manufacturing milk, developed in a well-structured framework, are necessary for assessing impacts of public policy for the dairy sector and for setting policy parameters to achieve desired performance levels for output, prices, producer income, and other indicators.

Management and Operations

The dairy processing and distribution industry, similar to other components of the modern agribusiness sector is now in the "information age." Sophisticated systems are used in assessing market conditions, organizing production and distribution operations, and managing stocks and supplies. Chapters 3, 7 11, and 12 address various aspects of the use of demand parameters in operations analysis for the dairy industry. Ault provides an overview, indicating from an industry perspective the increasingly important place for accurate estimates of demand parameters in management. He discusses areas of their direct use in operations analysis. The conclusion is that the efficient and successful firms in the dairy industry will increasingly depend on accurate short-term, intermediate term, and long-term demand parameters for scheduling production, and for decisions on purchases, market actions, and investment and market positioning.

A part of the operations analysis for these firms is product development and promotion. Cox, although as well addressing another theme for the book, provides a general conceptual analysis for assessment of advertising and demand. Issues of demand responses to advertising, product development, and market positioning are important from an industrywide viewpoint; for individual firms these same issues are relevant in choosing to differentiate products on the basis of characteristics identified in advertising campaigns. Advertising, product differentiation, sales enhancement and stability, and the balancing of market promotion with that of competitors are important for the private sector.

In Chapters 11 and 12, Oskam and Williams illustrate and evaluate uses of demand parameters in forecasting and in industry analysis, respectively. For forecasting, short-term responses to prices, seasonality in demand, and other fluctuations affect decisions on management of supplies, and more generally affect the organization of efficient distribution and production systems. Oskam shows how forecasting procedures have been integrated into the management decision processes for the dairy industry in the Netherlands. The integrated use of Bayesian estimation and demand forecasting techniques in scheduling and distribution systems for dairy production foretells an era of increased sophistication incorporating features of forecasted demand in operations analysis. Similar conclusions are drawn in Chapter 12 in William's review of the use of demand analysis in the dairy industry of the United Kingdom. Again, there is broad evidence that analysis will contribute to improved efficiency of operations in the production and distribution of subsectors of the dairy sector. Evidence also suggests a key role for demand parameters in dairy sector policy analysis.

Data and Empirical Issues

The measurement of demand parameters has received wide attention by the economics profession, and applications for the dairy industry are illustrative of problems in the estimation of consumer demand that are characteristic of many agricultural commodities. Measurement, empirical, and data issues are addressed in Chapters 4, 8, 9, and 10. Chapter 4 is concerned with demand theory and issues of estimation of demand parameters in systems contexts. For regulation of the dairy sector and for uses in operations analysis, own-price responses, income responses, and cross-price responses are increasingly important parameters. These parameters are highly dependent on the underlying structure employed for the estimation of the demand parameters. Systems methods provide a conceptually consistent approach for estimating and interpreting cross- and own-price effects as well as income parameters. Also, these systems contexts can be used for incorporating variables to reflect demographic, region specific, and other factors that impact demand responses. In modern applications, demand systems offer an opportunity for more appropriate merging of the theory of demand with the data, permitting tests of the theory and more accurate parameters for applied uses.

Haidacher assesses alternatives for applying demand theory in the estimation of parameters for the U.S. dairy industry. This analysis points to the current level of understanding of the structure of demand for dairy products and the role of systems techniques in providing an improved basis estimation of demand parameters. His chapter points up the value of understanding the limitations of estimated demand parameters and how interpretations of empirical results are conditioned by the data available and the estimation methods.

Schrimper (Chapter 8) and Buse (Chapter 9) raise a number of issues that are pervasive to dairy demand estimation and more generally to demand for agricultural products. Schrimper lays out a number of important issues for estimation demand systems that are raised by matching the theory with the applied requirements for demand information and the data systems. He concludes that demand systems must be highly specialized to the intended application and that estimation methods used in the development of these systems must recognize the underlying features of the empirical information employed.

Data sources and the quality of data systems continue as a limitation for the estimation of improved dairy demand parameters and demand systems. The review of data systems provided in Chapter 9 is for the United States. The situation is similar for other nations. The lesson learned from this analysis involves the limitations of the data sources and their information content relative to uses in the estimation of demand parameters. In general, the analysis shows that there are more sources of data than are being effectively used by demand analysts for estimating dairy sector parameters. And, unfortunately, demand analysts utilizing these data systems may not be as familiar as they should with the underlying assumptions and special limitations that derive from the way the data systems have been generated. Clear understandings of the data and the way these data have been generated and processed are necessary for the development and evaluation of dairy demand parameters.

ADVERTISING

Perhaps due to the large impact of producer-sponsored promotion programs in the United States, Canada, and other nations, there has been a flurry of activity in development of estimates of impacts of advertising on market demand. Chapters 6, 7, 10, 13, 14, and 15 address various aspects of the issues raised by linkages of advertising and demand. Chapters 6 and 7 lay out the conceptual basis, as it is possible, given the state of the art, for estimation of parameters designed to evaluate impacts of advertising on commodity demand. It is shown that the data available are quite limited for the task of estimation of the subtle impacts of advertising on demand. Among the problems involved in estimating promotion and advertising impacts is the fact that dairy products are already widely consumed by most populations in developed countries. Thus, the effect to be evaluated is one of the impact of increased or altered information, on the consumption of products already consumed at some level by most households/individuals. Reducing these behaviors to advertising information to shifts in demand functions involves the use of highly specialized assumptions on how individuals process information. These assumptions, and the theories on which they rest, are at present in the formative stages. The results discussed in Chapters 6 and 7 will lead the way to a clearer understanding of the conceptual foundations for estimation of advertising and promotion impacts.

Chapters 10 and 15 provide more background on issues of advertising and demand. Haidacher shows how the dairy demand in the United States has responded relative to prices and incomes. Much of the change in consumption patterns over time is shown to be related to price and income changes. The conclusion is that the potential for advertising impacts is limited and that specifications are necessary in advertising analysis that allow researchers to parcel out these potential impacts carefully. Hallberg illustrates the importance of linking market demand impacts to producer prices through the derived demand framework. These linkages are crucial for policy analysis and for study of promotion and advertising benefits from a producer perspective.

Chapters 13 and 14 contain results of estimates of advertising impacts for butter and margarine and fluid milk in Canada. These results and their assessments for the dairy industry should be viewed as preliminary. Secondary data or, more properly, the data at hand were employed in these analyses of advertising impacts. In general, the effects of advertising are shown to be positive or to increase or shift consumer demand. The conclusion is that consumer attitudes and perceptions and, accordingly, the expression of these attribute parameters describing demand responses can be identified with promotion programs for the dairy industry. These results on advertising impacts will require further refinement and assessment, using emerging frameworks like those suggested in Chapters 6 and 7 to more fully characterize advertising and impacts in a structural context.

CONCLUSION

This collection of issues, experiences, and results should be viewed as providing an overview and foundation for guiding the development and use of demand parameters for the dairy industry. The strong theoretical foundations explicated in this book can be used in the estimation and integration of dairy demand parameters. Statistical techniques, as well, have been developed that make it possible to efficiently utilize available data and nonsample information in improving the reliability of demand parameter estimates and in specializing these estimates to particular operational and policy uses.

Policy analysis and operational uses of dairy demand parameters are documented. From an operational viewpoint, it is apparent that there will be increased uses for dairy parameters in management and related firm-level decisions. The number of perspectives represented in this volume illustrate that timely and accurate information on market conditions can be effectively used in reducing processing and distribution costs and improving the efficiency of the processing and distribution industry. At the industry level, policies regulating the dairy industry are shown to be in a state of change. This is attributable to the multilateral trade negotiations, changes in the political economy for agriculture, changes in production and processing technology, and a host of other factors. As these policies are adjusted and restructured, issues of incidences and impacts of changes in policy will receive increased attention. Demand parameters are important elements in the calculation of these policy performance indicators and for the design and operation of policy and regulatory schemes.

Theory, methods, measurement, and data will occupy continuing positions of importance in dairy demand analysis. Data systems must be maintained and more fully understood as a basis for estimating improved demand parameters. Theories will develop that are more consistent with improved use of existing data systems. And, the data systems themselves will adapt. Still there will be specialized problems associated with the matching of data systems not completely oriented to the intented uses, and with specifications that are designed to produce demand parameters to be utilized in specific contexts. Evolutions of data systems, improved estimation methods, and adaptations of theory will be an ongoing feature of the application and use of demand concepts in operations and policy analysis for the dairy industry.

A new issue for the dairy industry highlighted in the book involves questions of advertising and promotion. Here, the conceptual basis for specifications used to evaluate advertising promotion and campaigns is addressed. Preliminary results of advertising evaluations for butter and fluid milk in Canada are presented. Significant funding is being used in the promotion and advertising of dairy products. Incidences of benefits of these advertising and promotion campaigns will be the source of continuing debate for the dairy industry. Conclusive results of course must await the development of improved theories on processing of information by consumers, the associated modifications in demand systems structures, and new data requirements suggested. In a sense, the advertising and promotion issue shows the importance of bringing together theory, estimation, and data for careful analyses of policy and operational problems in the dairy industry. Much is at stake for the processing and distribution firms, and for consumers and producers. Improved methods, new data systems, and results will follow. At the same time, the capacity of orienting existing frameworks to the assessment of advertising and promotion is illustrative of the value of demand parameters for the dairy industry. The significant contribution that careful analysis utilizing the most current theory and estimation methods and data systems can make to important industrywide policy issues is evident.

CHAPTER 2

Demand Parameters and Policy Analysis for the Dairy Sector

Garry R. Griffith, Ralph G. Lattimore, and John C. Robertson

THE MAINTENANCE of an adequate domestic dairy industry is believed to be important in most countries because fluid milk is regarded as a necessity, has a relatively inelastic demand, and until recently has been regarded as too bulky and perishable to make long-distance trade feasible. In fact, most countries have well-established dairy industries capable of satisfying domestic fluid milk requirements. The approach to dairy policy around the world is so prevalent that a general framework may be described to fit many situations. The aim in this chapter is to assess regulatory policies in the dairy industries. To add more realism, five particular dairy programs have been chosen to exemplify the issues and policy approaches.

The five major milk producing and exporting regions chosen—Australia, Canada, the European Community (EC), New Zealand (NZ), and the United States—all have used intervention measures to influence the domestic farm price for fluid milk (and milk products) to a level usually above what would have been determined by the market in the absence of intervention. For example, guaranteed minimum prices, quotas, price pooling, and buffer funds have all been applied, and in most cases these have been supplemented by export subsidies, import restrictions, input subsidies, and guaranteed purchases of excess production.

Most of these intervention measures have been aimed, at least in the first instance, at the supply side of the dairy industries. Pricing policies have been implemented to support and stabilize dairy farm incomes and/or to

The authors gratefully acknowledge the financial support of the U.S. Department of Agriculture (Cooperative Agreement No. 58-3J22-5-00202) and the New Zealand Ministry of Commerce.

maintain a stable supply of milk that will satisfy domestic market requirements for milk and milk products. Regulatory policies that impact on the farm level or manufacturing level price of milk or dairy products will have some effect on the prices charged final consumers for these products. Similarly, policies that impact more directly on the output of milk or dairy products will have an influence on the availability of these products to final consumers. In either case, consumers will face a distorted set of retail prices and quantities of these products, and their consumption decisions will be affected.

Interventionist dairy policies around the world over the last 50 years have also contributed to another phenomenon shared by the dairy markets—the prevalence of state trading. The degree of state trading and multinational involvement (often by state traders) in dairy markets has contributed toward a high propensity for tender and contract selling rather than transparent world market development as for grains and oilseeds.

DAIRY POLICY INTERVENTION MECHANISMS

The information in this section is taken from work in progress at Lincoln College (Robertson and Griffith 1988; Robertson et al. 1987a,b). Further results are available in those papers.

Nature of the Problem

Since World War II industrialized development has placed an increasing burden on the ability of the agricultural sector to maintain farm incomes comparable to those in the industrial sector.

During the 1950s and 1960s most of the regions selected for study had a relatively large proportion of small dairy farms and limited alternatives. In the European Community, the United States, and Canada this also created strong political demand for support. Advances in technology have removed a large component of the seasonality of milk production. Hence, price support/stabilization has come to be viewed as a means of maintaining the relative position of dairy farm incomes rather than as a means of stabilizing these incomes through the year. Administered or intervention prices in combination with supplementary measures such as quotas, buffer funds, price pools, and input subsidies have formed the basis for domestic milk pricing and have been the key instruments of income support programs. However, among the industrialized dairying regions there exists a wide variation in the price-setting mechanisms and their responsiveness to world market prices and trade conditions. Overall though, in most developed countries, pricing policy has been the main way by which governments have stimulated or discouraged milk production as well as influenced shifts in income transfers.

The close interrelationship between these production and support policies becomes apparent if examined within an *administered pricing framework*. Administered or support pricing describes the situation in which an organization, be it a producer authority or central government, sets guaranteed prices for farm products via regulation.

Typically, the objective of administered pricing in dairying has been to meet the income goals of the industry, balancing supply and demand in a manner that achieves desired income transfers and does not produce significant government expenditures. However, artificially setting prices perceived adequate to maintain farm incomes relative to nonfarm incomes usually promotes production in excess of what can be sold "efficiently" to satisfy domestic and export demand. That is, the price support not only maintains farm incomes but also stimulates investment, technical advance, and, consequently, output. Supply response studies indicate milk production is relatively responsive to changes in price in the medium term (Salathe et al. 1982).

An illustrative model of administered pricing for a dairy product is provided in Figure 2.1. Supply of the dairy product is given by S and corresponding demand by D. The world price, P_w, is the global market clearing price for the product. In this example the region would be a net exporter in the absence of domestic price intervention. However, in an administered pricing situation the internal price P_i is set by fiat. At this price domestic supply exceeds domestic demand by $(S_o - D_o)$. If excess supply is exported in the current period at the world price, then the value of necessary export subsidies is represented by the hatched area. Alternatively, a proportion of this excess supply may be stored in expectation of a future rise in the world price.

Of course, the resulting excess production must be financially supported if incomes are to be maintained in the future. McCalla and Josling (1985) suggest a number of common approaches to finance-administered prices. These include the taxation of consumers, and a reduction in domestic demand; direct government transfers such as subsidizing feed costs; import controls; and export assistance. For example, net U.S. government expenditures for dairy in 1983 were around U.S. $2.6 billion. In the EC for the corresponding period, 4.7 billion European Currency Units (ECU) was spent. Thus, the support mechanism of guaranteed purchases of milk (usually in the form of butter and skim milk powder) by intervention agencies in the United States and the EC, initially designed as a "safety net" for producers, has become an outlet in its own right.

Transfers from taxpayers and consumers can be limited by supply quotas in a variety of ways. For example, the Canadian dairy program has

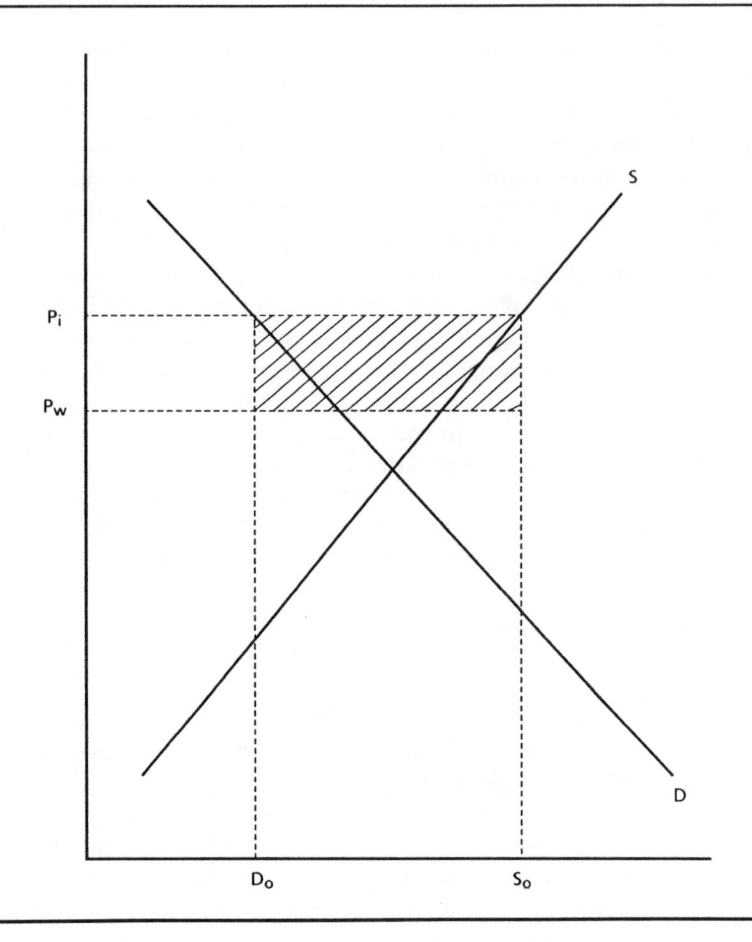

Figure 2.1. General representation of an administered price regional commodity model

explicit supply management quotas that limit the payment of P_i to a quantity of output less than S_o but greater than D_o.

Each of the representative countries has adopted some variant of this administered pricing model.

New Zealand. In international trade-dependent regions, world prices and thus export returns tend to be dominant influences of the domestic returns to producers. New Zealand exports approximately 80 percent of milk production.

The influence of the world market on the New Zealand dairy industry is through the prices offered to processors for all milk production destined for export. The sole exporting authority, the New Zealand Dairy Board, sets and administers these offer-to-purchase prices in advance which in turn imply guaranteed returns to milk producers. The offer-to-purchase prices are the New Zealand intervention prices (P_i) in terms of Figure 2.1. They may be higher or lower than the world market price that eventuates.

However, since expected and realized exports are different, the pricing system must be maintained from accumulated savings or borrowing by the dairy board. Thus, instability in the world market is accommodated by a domestic buffer fund or stabilization system. In periods of high world prices relative to the guaranteed prices, one-half of the export trading surplus is withheld from producers. The proceeds are then used to augment returns to producers in periods of low world prices. Offer-to-purchase prices at least partially insulate domestic producers from deviations in expected world prices. As long as surpluses balance deficits, little financial strain is placed on existing resources, the system can be industry funded, and prices can be set to insulate producers from world market fluctuations (Clough and Isermeyer 1985).

Historically, guaranteed prices for dairy products were set in close accord with expected world prices. However, a successful buffer-fund system requires that two related conditions hold: first, that nominal world prices, expressed in New Zealand dollars, show at least an overall stable positive balance relative to costs; and second, that periods of declining or lower world prices are sufficiently short run. These two conditions dictate the degree to which administered prices in any year relate to the actual world price.

Because the New Zealand currency has tended to depreciate in value since the mid-1970s relative to that of the other regions, the effects of falls in the world prices have been dampened and similarly, rises in world prices have been amplified. For the period 1960 to 1986 the balance of the trading surplus/deficit account of the dairy board stabilization account has tended to fluctuate around zero with few periods of sustained deficit or credit.

Until 1985 low interest cost financing was available from the Reserve Bank of New Zealand. This low interest rate was paid by the bank on credits in the dairy stabilization accounts. Excessive use of this concession was generally not required because adequate returns were being made from international trade. But the financial loss of the dairy board for 1986 is a clear example of the consequences of setting administered prices significantly out of line with world prices when operating an essentially self-financing buffer fund.

In addition, a number of other direct influences on the profitability of milk production have been exercised recently by the New Zealand government. These have included subsidies to reduce production costs, interest rate subsidies, subsidies to expand domestic demand, and in one year the use of a deficiency payment. However, recent estimates by Johnston (1985) indicate that the effects of these subsidies have been minimal. For the period 1977 to 1981 the New Zealand dairy subsidy per liter of milk was estimated at less than NZ$0.008. Moreover, recent changes in industry assistance policy in New Zealand have seen the removal of almost all the subsidies referred to above.

Price and production quotas for fluid milk were set separately by the New Zealand Milk Board until 1987. From 1987 the fluid market arrangements have been considerably relaxed. Only 8 to 10 percent of milk production is consumed domestically in fluid form.

Australia. Production destined for the fluid and manufacturing milk markets is carried out on most dairy farms. Also, as in the United States and Canada, Australian pricing and quota policies for the fluid market (which affect above 25 percent of milk production) are set separately from the industrial milk sector of which around 75 percent is exported. However, it is the policies for industrial grade milk production that provide the underlying base for the dairy industry, since they tend to set minimum market conditions for milk production and trade. In Australia, single statutory producer authorities have dominated the industry's history.

There have been a number of major changes of policy in Australian dairying over the last twenty years (Griffith et al. 1988). To summarize, policies have tended to be introduced as five-year plans. During the period up to 1967 a voluntary equalization scheme operated that provided common, pooled returns to producers of particular milk products irrespective of whether they were sold on the domestic or export markets. Since 1962 these equalized returns have been underwritten by the federal government and supplemented by a system of processor bounties for certain high-fat products (including those for export).

From 1967 to 1972 the equalization scheme was continued but supplemented by government devaluation compensation payments in response to devaluation of sterling vis-à-vis the U.S. dollar. From 1972 to 1974 the underwriting of producer returns was abolished and bounty arrangements were phased out. In 1975 underwriting was reintroduced for skim milk powder and in the following year extended to include butter and cheese.

From 1977 to 1986 levies were imposed on certain milk products sold on the domestic market to maintain profitability of exporting at the expense of reduced domestic demand. Export returns were then pooled and equalized to provide a high farm-gate price, which encouraged further production discouraging resources flowing out of the dairy sector. To supplement these new pooling and equalization arrangements, underwritten prices were again set. Export incentives in the form of grants for market development and export expansion were also used to assist small- to medium-sized processing firms.

Beginning in 1986, measures to significantly reduce the level of support for the dairy industry were implemented. The aim of these new arrangements has been to provide clearer price signals to producers and allow decisions more responsive to world market conditions. The price support is now funded by a levy on all milk production, the system of pooling within product classes has been abolished, and domestic prices have been reduced to a level based on landed New Zealand export prices. The degree of success of these measures is still to be seen. All subsidies have not been removed, and import controls or voluntary restraints remain a significant feature of the Australian dairy industry. Moreover, recent interstate disputes, with low-cost industrial milk producers in Victoria undercutting their New South Wales counterparts, raise questions concerning conflicting interest between federal policy aims and individual state concerns with immediate producer needs.

In summary, the Australian dairy industry has historically been subject to a number of output subsidies including bounties, devaluation assistance, export incentives, price pooling and equalization, consumer-to-producer transfer payments, and underwriting schemes. As in New Zealand a number of general agricultural and input subsidies have been used to support the dairy industry. These include input subsidies on fertilizer and cattle health, extension services, concessional credit, income tax relief, and research. Johnston (1985) estimates that the total of these subsidies provided a direct benefit to Australian dairy producers of between $A0.05 and $A0.08 per liter of whole milk for the period 1977 to 1981.

The United States. In the United States the domestic market is viewed as the main outlet for milk production. Also, because of the relatively high concentration of small dairy farms, pricing policies have tended to be oriented to maintain real farm incomes. Thus, the policy has been set by the degree to which administered prices must be raised to achieve targeted incomes that has been the primary concern.

An additional factor influencing the level of support in these regions is the predominant use of high-cost concentrate feed inputs. Although these inputs produce relatively higher milk yields than for the pasture-based industries in New Zealand or Australia, the Australian and New Zealand industries are significantly lower cost.

Since 1949 the U.S. federal government has had significant involvement in the pricing and marketing of milk (USDA 1984). There are three basic elements in the federal dairy program: (1) a guaranteed administered minimum price for milk; (2) support for this guaranteed price through government purchases and disposals (including export) of milk products (cheese, butter, and skim milk powder) at announced prices; and (3) the use of import quotas. In addition, there has been a complex system of marketing orders isolating regional fluid milk markets.

Until the early 1980s the administered milk price was tied to a real price index set in 1910–14 (new parity price). The aim was to meet domestic market requirements, reflect changes in production costs, and assure a level of farm income sufficient to maintain production capacity into the future at a fair price plus a reasonable profit. The support price ranged between 75 and 90 percent of parity until 1980, and there was no direct link between the intervention price and the world market price.

During the 1960s the United States was principally a net importer of dairy products, and until the late 1970s domestic demand in the United States by and large absorbed the supply of manufacturing milk. However, over the last ten years the combination of rapid increases in production, high intervention prices along with unstable and falling world prices necessitated increased direct government intervention in the market through increased purchases of excess dairy production. The Commodity Credit Corporation purchased about US $2 billion in dairy products each year during 1981–86 and has accumulated stocks of nearly 20 billion pounds of milk equivalent product (Gardner 1987).

The cumulative effect of maintaining high guaranteed domestic prices relative to world prices during this period resulted in strict controls being placed on the rate of growth in support prices in the early 1980s. In 1981 legislation was passed that saw support prices being specified in dollar rather than parity percentage terms. However, production continued to expand and legislation in 1982 froze support prices for two years. Subsequently, support prices have decreased each year. Production controls were introduced in the form of a milk limitation scheme for 1984 and a system for the buy-out of whole herds was introduced during 1986. As a consequence of the 1985 Farm Bill (Food Security Act) the support price for milk is to be directly linked to the level of overproduction (as indicated by the level of government market removals into storage and for export) from 1987.

Although the growth in disparity between the U.S. guaranteed milk price and world prices for milk products appears to have been stabilized as a result of these measures, recent estimates still place the economic costs of the dairy program at around US $330 million per year in 1980 dollars (LaFrance and de Gorter 1985).

The European Community. The European Community (EC) is one of the world's largest milk producers and the largest exporter. Some member states are traditionally exporters, others importers. The EC dairy industry operates under a similar system of intervention to that in the United States.

However, in the EC the level of support has stimulated overproduction to such an extent that since the early 1970s large residual stockpiles of dairy products have accumulated. The support provided dairy production in the EC is afforded by intervention in two areas—intervention in trade between the EC and the rest of the world (variable levies on imports and export "restitutions" or subsidies) and intervention in the domestic wholesale market through the purchase of excess milk production by intervention agencies at guaranteed prices that will yield a target (and consequently market minimum) price for raw milk above the world price (see also Farber et al. 1984). As in the United States, the benefits of this type of public support go largely to producers in direct proportion to output and have no direct relation to the producer's individual income. Unlike the United States, however, where government intervention is a last resort, the EC plans expenditures in advance to cover anticipated excess production. This planned expenditure is calculated to cover the intervention price for expected production and associated excess supply.

In a market where there are substantial supply-demand imbalances, the intervention price ultimately determines the market price and, therefore, the price paid to producers. As a barrier to the importation of milk products at lower world prices, a threshold price is applied at the EC border. Imports of butter and cheese from New Zealand are permitted, but up to a specified quota based on special conditions. Excess production must be sold on the world market at a loss with the difference made up by an export subsidy (restitution) payment.

In response to both the fiscal and economic costs associated with these policies a number of measures have been introduced over recent years to control overproduction. However, these measures have met with only limited success. A possible reason for the seeming inability to secure immediate and significant reduction in production lies with the EC political system, which requires unanimous approval on these types of issues. This unanimity is very difficult to achieve given the diversity and complexity of the socioeconomic and political environments that exist in the EC. Since 1977 the EC has operated a system of coresponsibility levies on milk producers in order to indirectly lower prices and reduce the cost of the disposal of excess production. However, the levy has been set at only between 0.5 percent and 3 percent of the target milk price while the proportion of the annual EC agricultural budget being appropriated to dairying has remained at around 30 percent (Burrell 1987). Also, total and partial exemption from the levy was granted to some farmers in "less-favored areas."

In 1984 a "super levy," or quota system, was introduced to reduce milk production to the 1981–83 average level for the next five years. The hoped-for effectiveness of the quota has been largely reduced, however, because of the ability of processing companies to average out the levy over all suppliers and even allow interregional transfers of quota. The net result has been that in some instances increased milk production is still economically viable. In fact, milk production actually rose immediately following implementation of the quota levy.

Over a period of time these measures for limiting supply have had some impact, and there are indications that milk production in the EC has peaked and is making a steady if slow return toward some degree of domestic supply-demand balance. However, estimates of the duration of this restructuring process do little to alleviate the concerns of other exporting nations about the disposal of accumulated stocks of dairy products in the meantime.

The basic idea of the traditional U.S. and EC administered pricing arrangements using guaranteed government purchases of dairy products is that the intervention, or support price, is set at a level higher than the market clearing world price. In the absence of a favorable shift in demand or equivalent effective supply control, the net result is higher-than-otherwise consumer prices and producer returns and a residual output that must be exported at a subsidized price or stored.

Canada. In Canada the milk and milk products industry is regulated by both federal and provincial government policies. These policies include recommending milk prices, supporting the prices for dairy products, administering producer quotas to limit production, distributing direct subsidy payments to producers, and undertaking certain trade and marketing functions (Agriculture Canada 1983). These policies aim to ensure the domestic requirements for milk and milk products are met from domestic production without accumulating large stocks of processed products, and to provide a stable real return to producers. However, declines in domestic requirements and growing international surpluses have placed an increasing strain on the Canadian dairy industry's ability to meet these objectives.

The successful implementation of dairy policy in Canada requires a delicate balancing or sequencing of management decisions (Stonehouse 1979; Barichello 1981) as follows:

1. The support prices for butter and skim milk are set at levels to ensure a guaranteed market price. Since the mid-1970s this guaranteed price has been formulated solely on the basis of production-cost and consumer-price indices. Direct subsidy levels are also selected to supplement the guaranteed market price to yield a target producer price.
2. Domestic requirements for milkfat are estimated based on this

support price.

3. Some combination of quota levels and penalties must be selected to induce the correct supply of milk to satisfy domestic demands. These quotas must be distributed among individual producers.
4. Supply management is organized on the basis of providing supply-demand balance in milkfat, not whole milk. Solid nonfat or skim milk production is usually in excess of domestic requirements. Nonfat levies are paid by producers to help fund the export of these surpluses at a loss because world prices are typically lower than domestic support prices.
5. Foreign competition must be controlled or excluded. Import tariffs apply to all dairy products except butter, supplemented since 1951 by import quotas.

As a consequence of these price support and supply control measures, Canadian consumers face a high domestic price, taxpayers must fund the direct subsidies, and producers (in particular large producers) are guaranteed a high return for their restricted supply. There are substantial income transfers under the system. Barichello (1981) estimates that dairy producers benefit from an income transfer of about C $670 million per year, consumers incur an income loss of C $690 million, and taxpayers pay around C $300 million to finance the direct subsidies and to fund the disposal of excess skim milk production not covered by the nonfat levies. Barichello also notes that overseas consumers of Canada's surplus milk products benefit from being able to purchase those quantities at below their costs. The resulting net loss in social welfare under the current dairy industry policy was estimated as at least C $200 million per year (Short 1986).

In the Canadian system, supply is restricted to a maximum level determined by the quota. Production in excess of quota is uneconomic because of prohibitive overquota levies. Reduced levels of quota below the supply-demand balance make the support prices ineffective and raise product prices, while raised quota levels yield surpluses (in both skim milk powder and butter) and lead to increases in subsidized disposal on world markets.

All of the major dairy producing or trading regions considered intervene in the market for milk and milk products. This intervention has been principally in the form of guaranteed or support prices for milk. In New Zealand, the emphasis appears to be largely on income stability with intervention used principally to coordinate the price formation process. As a consequence of low production costs the industry can operate effectively with little additional support. In Australia a number of policies have been implemented based on guaranteed equalized prices and subsidies. These measures have caused some misallocation of resources and economic welfare losses. In the United States and European Community, policies have stimulated high overproduction and the income transfers involved have become extremely large. High production costs in these regions have necessitated high levels of support to maintain the industries. Canada has sustained a certain degree of domestic supply-demand balance in milk but has done so at the expense of consumer taxes, high consumer prices, and taxpayer transfers.

Overall, it is clear that the principal beneficiaries of intervention are the domestic milk producers and consumers in the rest of the world. The cost of intervention is borne primarily by domestic consumers and taxpayers. The relative levels of these economic costs and benefits depend on the level of association between the prices offered for domestic milk production and the underlying world market conditions. A summary of the various policy instruments implemented in each region is provided in Table 2.1.

POLICY INTERVENTION MECHANISMS AND THE DEMAND FOR MILK AND DAIRY PRODUCTS

As discussed for the selected regions, administered pricing systems have been implemented to maintain and stabilize dairy farm incomes, to ensure an adequate supply of milk, and to finance income transfers. The demand for milk and dairy products is impacted through either the price paid by consumers for these products or the quantities available for purchase by consumers. A number of these possible impacts are outlined below.

High Retail Prices

To be able to offer relatively high administered prices to dairy producers and/or to finance the expenditure subsidy required, retail prices for milk and dairy products in most regions are set above competitive market levels. Thus, domestic demand for these products is below the level in the absence of intervention. Retail prices are maintained above the intervention prices by the imposition of fixed distribution margins and by regulated minimum retail prices. However, as shown in Table 2.2, most own-price elasticities of demand for dairy products are in the inelastic range, so continual increases in retail prices result in proportionately lower declines in consumption and proportionately greater income transfers.

Table 2.1. Dairy product policy instruments in New Zealand, Australia, United States, European Community, and Canada

	Region				
Policy instrument	New Zealand	Australia	United States	Canada	European Community
Milk production controls	*[a]	*[a]	*	*	*
Administered/ guaranteed farm prices	*[b]	*	*	*	*
Administered/ guaranteed product prices	*[b]	*	*	*	*
State trading	*	*		*	
Import controls		*	*	*	*
Government purchases of surplus			*		*
Export subsidies			*	*	*
Administered retail prices	*	*		*	
Promotion/demand subsidies/restrictions on substitutes	*	*	*	*	*

Source: OECD (1987).
[a] Fluid milk sector only.
[b] Assisted by a buffer fund.

Stable Retail Prices

The administered price policies also have stability objectives. Thus, constant or relatively smoothly increasing nominal intervention prices are set to ensure stable incomes to producers and stable supplies of fluid milk to consumers. This history of relative stability in retail prices has possibly induced the inelastic price response apparent in these products and led to a greater influence of tastes and preference in the demand for these products. For example, in most regions per capita consumption of butter has declined steadily over the past two decades as attitudes toward dietary fats have altered (Griffith et al. 1988; Robertson et al. 1987a).

Table 2.2. Typical dairy product price elasticities of demand for New Zealand, Australia, United States, European Community, and Canada

	Elasticity with respect to the price of						
	Butter	Cheese	Whole milk powder	Condensed milk	Skim milk	Fluid milk	Other milk
Demand in New Zealand							
Butter	−0.5						
Cheese		−0.5					
Whole milk powder			−0.5				
Condensed milk				−0.5			
Skim milk					−0.5		
Fluid milk						−0.5	
Other milk							−0.5
Demand in Australia							
Butter	−0.5						
Cheese		−0.13					
Whole milk powder			−0.45				
Condensed milk				−0.45			
Skim milk					+0.45		
Fluid milk						−0.3	
Other milk							−0.45
Demand in United States							
Butter	−0.56						
Cheese		−0.60					
Whole milk powder			−0.10				
Condensed milk				−0.84		0.84	
Skim milk					−1.72	1.72	
Fluid milk					0.10	−0.32	
Other milk							−0.34
Demand in European Community							
Butter	−0.43						
Cheese		−0.30					
Whole milk powder			−0.10				
Condensed milk				−0.85		0.60	
Skim milk					−0.30	0.20	
Fluid milk						−0.19	
Other milk							−0.33
Demand in Canada							
Butter	−0.86						
Cheese		−0.91					
Whole milk powder			−0.33				
Condensed milk				−0.33			
Skim milk					−0.19	0.09	
Fluid milk						−0.44	
Other milk							−0.33

Source: Policy Services Division, MAFCORP, Wellington, New Zealand.

Restrictions on Domestic Supplies

Many regions directly restrict the quantities of fluid milk and milk products available for consumption. Fluid milk production quotas are implemented in New Zealand, Australia, the European Community, and Canada and have the effect of restricting the availability of supplies for domestic consumption. As mentioned above, this is one method of reducing the taxpayer commitment to the administered price system.

Restrictions on Imports

Another method of maintaining administered price regimes is to restrict imports quantitatively through quotas (Australia, United States, European Community, Canada) supplemented by import tariffs, which raise the price of imported product to the domestic intervention price (United States, European Community, Canada). Again, such policies restrict the quantity and/or variety of products available to domestic consumers and force them to pay prices above those current in the world market.

Promotion Programs

All regions have implemented promotion campaigns to induce domestic consumers to purchase more of the domestically produced product. Such programs aim to expand domestic demand and reduce the quantities of surplus milk products that have to be financed under the administratively set support price schemes. However, with low income elasticities (except for cheese), seemingly well-entrenched attitudes toward milk products, and in general increasing real domestic prices, per capita demand for milk products has tended to continue to decline in the major producing regions.

Restrictions on Substitutes

Another way of forcing domestic consumers to purchase milk and milk products in the face of high and increasing retail prices is to restrict possible substitute products. Thus, for example, in New Zealand, Australia, and Canada up until the mid-1970s the sale of margarine was either banned or restricted by very small production quotas and regulated by a variety of labeling and packaging restrictions that benefitted the dairy industries. In New Zealand, when the ban on margarine was lifted, the retail price of butter was subsidized for several years to maintain a price advantage.

In winter months, processors in some countries are tempted to provide reconstituted milk as a substitute for fresh fluid milk. Regulations often restrict or prohibit this competition to encourage winter milk production.

State Trading

It is not unusual for international markets in agricultural products to be influenced by a small number of large multinational trading firms. This is true of international dairy markets except that often the agents are state trading firms that emanated from national producer cooperatives.

During the 1920s and 1930s a part of the response to low dairy farm incomes involved stimulating dairy producer cooperatives in countries like Australia, Canada, France, New Zealand, and the Federal Republic of Germany. As administered dairy pricing regimes developed, so did the size and influence of these cooperatives/state traders.

Today, organizations like the Australian Dairy Corporation, the Canadian Dairy Commission, and the New Zealand Dairy Board are multinational concerns involved in processing and trading dairy products outside their national borders. It can be argued that this growth in state trading has been stimulated by the lack of world dairy markets (i.e., markets exposed to world trading conditions). The presence of these organizations has also stimulated some degree of world market coordination such as the GATT dairy minimum pricing arrangement. Such arrangements, however, have not yet been formalized into institutions like the International Wheat Council in the grains area. The cooperatives are important alongside private multinationals in transferring technology at the farm and processing level, especially to less developed countries.

Overview

Examples have been given to show how government intervention in the dairy industries of the major producing and trading regions affects the demand for milk and milk products. Administered pricing schemes employed to maintain and stabilize dairy farm incomes, to ensure adequate and stable supplies of milk, and to finance program expenditures have substantial impacts on the prices faced by final consumers and the availability of products. Effectively, the milk overproduction–farm income problem common in these regions has been transferred to domestic consumers, taxpayers, and foreign producers.

The challenge is to measure empirically some of these impacts of the administered pricing systems on consumers.

REFERENCES

Agriculture Canada. 1983. *Food and Agricultural Regional Model-Dairy.* Ottawa, Ontario: Information Services, Agriculture Canada.

Barichello, Richard. 1981. *The Economics of Canadian Dairy Industry Regulation.* Ottawa, Ontario: Economic Council of Canada and the Institute for Research on Public Policy.

Burrell, Alison. 1987. "EC Agricultural Surpluses and Budget Control." *Journal of Agricultural Economics* 38:1–14.

Clough, P. W. J. and F. Isermeyer. 1985. *A Study of the Dairy Industries and Policies of West Germany and New Zealand.* Agricultural Policy Paper No. 10. Palmerston North, New Zealand: Centre for Agricultural Policy Studies, Massey University.

Farber, U., K. Frohberg, E. Geyskens, C. Meghir, and P. Pierani. 1984. *The IIASA Food and Agricultural Model for the EC: An Overview.* IIASA Working Paper 84-50. Laxenburg, Austria: International Institute of Applied Systems Analysis.

Gardner, Bruce L. 1987. *The Economics of Agricultural Policies.* New York: Macmillan.

Griffith, G. R., J. C. Robertson, and R. G. Lattimore. 1988. *An Econometric Model of Australian Dairy Industry Policy, 1965–1986.* Research workpaper. Sydney, Australia: N.S.W. Department of Agriculture.

Johnston, P. V. 1985. *Australian and New Zealand Dairy Industry Programs.* International Economics Division Report No. AGES841228. Washington, D.C.: USDA, ERS.

LaFrance, J. T. and H. de Gorter. 1985. "Regulation in a Dynamic Market: The U.S. Dairy Industry." *American Journal of Agricultural Economics* 67:821–832.

McCalla, A. F. and T. E. Josling. 1985. *Agricultural Policies and World Markets.* New York: Macmillian.

OECD. 1987. *National Policies and Agricultural Trade.* Paris: Organizaiton for Economic Co-operation and Development.

Robertson, J. C. and G. R. Griffith. 1988. "An International Comparison of Dairy Support Policy." Unpublished paper. Lincoln, New Zealand: Lincoln College.

Robertson, J. C., G. R. Griffith, and R. G. Lattimore. 1987a. "Dairy Product Policy in New Zealand 1965–1985: An Econometric Analysis." Adelaide, Australia, AAES conference paper (February).

Robertson, J. C., G. R. Griffith, and R. G. Lattimore. 1987b. "An Econometric Model of New Zealand Milk Production and Utilization." Blenheim, New Zealand, AAES (New Zealand Branch) conference paper (July).

Salathe, L. E., J. M. Price, and K. E. Gadson. 1982. "The Food and Agricultural Policy Simulator: The Dairy-sector Submodel." *Agricultural Economics Research.* Washington, D.C.: USDA.

Short, C. 1986. "Reducing the Cost of the Dairy Program." *Canadian Journal of Agricultural Economics* 34:379–397.

Stonehouse, D. P. 1979. "Government Policy for the Canadian Dairy Industry." *Canadian Farm Economics* 14:1–11.

United States Department of Agriculture. 1984. *Dairy: Background for 1985 Farm Legislation.* Agriculture Information Bulletin Number 474. Washington, D.C.: USDA, ERS.

28 - 45

CHAPTER 3

Demand Parameters and Operations Analysis in the Dairy Industry

Q 11

Donald E. Ault

ECONOMETRIC MODELING is relatively new to business decision makers. Most applications have occurred during the past 20 years. And, their acceptance has been slow and their use somewhat limited. Nevertheless, business analysis and decision making have been improved by the use of these analytical models according to most industry users. The range of inquiry has been expanded and a degree of precision not possible from industry knowledge alone has evolved. When used to complement industry experience and knowledge, models and statistical parameters become a powerful means of addressing alternative decision possibilities.

USES IN MANAGEMENT

Applications of demand models in domestic business operations range from simple sales forecasts to rigorous commodity market analyses to complex and comprehensive policy evaluations. The range of application is largely dependent upon the skills and preferences of analysts and other end users. Some business environments may exhibit biases against particular quantitative approaches. More often, the focus is on meaningful results and plausible explanations rather than particular techniques or tools utilized. Practitioners usually apply the analytical method that they believe is best equipped for the problem at hand. The growing capability of models has enhanced their application considerably.

Many business decision environments dictate unusual care and prudence in model usage. Careful use of the analytical device is necessary to avoid jeopardizing its future use as a decision aid. Prudence on the part of the user requires communicating the limitations of the approach along with its value and usefulness. Overselling the technique and the underperformance of results can damage the future acceptance of decision approaches that rely heavily upon the analytical technique.

The applications of models in dairy product export markets are severely constrained by the protectionist trade policies and limited commercial trade pursued by nearly all major milk-producing countries. Under a more open trading environment, considerable attention would be devoted to discerning reliable demand elasticities of the respective domestic and export markets. The associated models would be used to a greater degree to develop pricing strategies maximizing revenues. Countries utilizing dairy boards or centralized marketing agencies can make use of these demand parameters as they routinely differentiate prices. However, it is more common to use the export market for surplus disposal and, then, price becomes incidental to the overall marketing strategy.

The possibility of freer trade in dairy has encouraged the World Bank, USDA, and others to model the global market to determine the impact on consumption and price of liberalized trade. These efforts provide valuable insights for policymakers and trade negotiators. Additionally, they provide the basis for long-term business plans involving company policy positions and market development efforts. Capital in business is deployed cautiously on the basis of potential policy changes. But efforts are made to expedite the policy change if there is knowledge that market opportunity exists. Conversely, if a loss of market or competitive position appears imminent, many businesses will direct capital expenditures and marketing efforts elsewhere, positioning for greater opportunities in other ventures.

Another growing use of demand parameters in dairy is connected with the expenditure of large amounts of promotional monies. Allocation of funds can be driven by parameters indicating how maximum leverage from limited promotional dollars can be achieved. Promotional checkoffs generate large funds that can be used for generic advertising, featuring, promotional allowances, and research on a multitude of milk products. The promotional strategy can be further targeted toward age groups or classes of consumers, income strata, geographic regions, and specific markets and uses. Demand parameters are useful in helping make these choices. Far too often, however, these allocation decisions are made on less than adequate empirical data. Frequently they are made based on intuition, judgment, and much less sophisticated and reliable analytical approaches.

Branded marketing strategies within private companies may utilize similar demand parameters. However, data limitations and inadequate analysis or lack of market segmentation may leave the marketer often with only experience and judgment to support allocation decisions. The more astute marketers will develop measurement devices to aid in these allocation decisions. At the very least, trial and error approaches can be deployed. Sometimes habit or custom and what "appears to work" becomes the guide. This may be a very successful strategy and one that has been proven to the decision makers' satisfaction.

One of the serious problems that business must deal with in decision making is the multitude of demand parameters generated by research and analysis. The data source, the time period examined, and the particular model selected often yield quite different parameters, as the researcher would expect. However, the business decision maker often finds these divergent results confusing and misleading. Business logic would dictate a close convergence of parameters, but research often does not yield this conformity of results. These differences can lead to frustration and dismay on the part of many decision makers. Unless explained and communicated effectively by the researcher, decision makers will discount the value of the approach and technique used.

Early Experience with Demand Models

Early experience with industry models and demand parameters made business people cautious and skeptical about their widespread use in business operations. Often the technique and parameters developed were oversold and underexplained. Limitations were not adequately communicated. Familiarity with the tool and its potential often superseded understanding of the industry and the particular needs of the decision maker. This failure to adequately understand the uniqueness of individual industries or product lines resulted in mis-specifications, overlooking key aspects of the problem area addressed.

In a corollary sense, early econometric demand models were like the new food product that research and development scientists developed for production and marketing, which did not live up to early promises. It met the markets needs, it was top quality, it was a new concept—nutritious, tasteful, full of desirable attributes—and it had the makings of a winner. However, in test market, while initial acceptance was good, the repeat business was a problem. Why? An oversimplified but accurate diagnosis would be that it was improperly promoted, often misused, misunderstood, poorly positioned to the end user, and performed below expectations. In short, it was a tough sell. As in food products, the product itself is often not the problem, it is the positioning of the product, pricing, competition, merchandising strategy, and most important of all, the customer orientation.

INTEGRATING QUANTITATIVE MODELS AND JUDGMENT

There is little question about the value of demand models and how they can aid decision making. The problem is in their application, explanation, and the education of the end user to make the results meaningful and useful in a business or operational sense. Until most businesspeople become adept at understanding demand models, including their uses and limitations, modelers will need to focus more on applications and proper communication of results.

Early development and application of demand models in business seldom recognized the value of the intangibles: judgment, experience, and intuition. This was unfortunate because it led to distrust and delayed the success for demand models in many possible applications. Furthermore, it led to a communication and application gap that became more difficult to bridge within the ranks of business. This delayed and often prevented the full integration of modeling with operational decision making. In some cases it isolated the skilled practitioner in the decision hierarchy to a mere "input provider" role and an occasional advisory resource. On other occasions the modeler would become the specialist who was perceived as technically capable of performing the specialized role of running the models but not of applying the results to aid in making actual decisions. As a result, some of the synergistic possibilities for modeling and decision making were never adequately explored.

DECISION MAKER NEEDS

The needs of business decision makers are quite diverse for decision-making aids. First, the types of decisions by different levels of management whether by marketers or operations people are often quite different. These decisions in policy formation and developing company positions on various public policy issues will also be different. But there are a number of desirable attributes that businesspeople want from these decision aids. Included among them are the following:

- Problem orientation
- Simplicity
- Flexibility
- Accuracy and timeliness
- Understandability

Assistance in problem orientation and formulation is one of the more essential attributes for a management decision tool. Properly defined problems and tools that aid in that process are needed to adequately explore and resolve the decision options satisfactorily. Often a problem may be incorrectly defined and because of it the possible solutions may not satisfy the decision maker's needs. Clearly, agreement must be reached between researcher and decision maker on the specific problem addressed and the dimensions of that problem.

Communications and acceptability of quantitative models often depend on how well end users follow the logic of the approach. For this reason simplicity is preferable to complexity. This does not mean the analytical tool or the problem formulation need to be unnecessarily simplified, but the explanation of the approach, internal logic, and results are better received if kept simple. Subsequent and repeated use of the results may depend upon the relative ease with which the decision maker can communicate the results to others.

The more often a model or its results can be adapted to different problem situations the more valuable it becomes as a decision aid. Flexibility in addressing a number of possible situations or environments makes it more likely that the demand model will become a regular decision aid. If not adaptable to different sets of conditions it becomes less useful and overly specialized for widespread use. An example would be a demand model that yields both good elasticity or price flexibility estimates and quality forecasts of market price.

Decision makers want and expect accuracy from models in order to rely upon them in key decision situations. They also expect results and solutions to be available at the time needed for decisions. If the results are not available at that time the model has little value from a decision-making standpoint. Therefore, the need for easy to use and easy to update models cannot be overemphasized. This may require in-house developed models as opposed to those developed by outside researchers.

The correlation between use of a technique or model and understandability is high in business. Seldom are approaches or techniques used that are not clearly understood by decision makers. As in most instances, people prefer the familiar to the unknown and will use what they understand, can explain, and have confidence in. The message to the model builder and practitioner is to improve the understanding of the process, its rationale, and, above all, the logic of the results. In many cases this requires involvement of key participants in the decision process prior to sharing of the final results.

These five attributes of a model and the model user will advance their use and help in insuring favorable acceptance by business. The criteria are difficult and exacting and perhaps cannot be achieved completely or continuously. However, attention to these factors, along with the individual decision maker's unique needs and motivations, will go a long way toward gaining greater acceptance of quantitative approaches.

HISTORY OF HOW MODELS ARE USED

Agribusiness has used various econometric based models for approximately 20 years. Their early use in the late 1960s was very limited. Few industry economists were trained to be adept in their development or sufficiently skilled in applying them to real business problems. Use was limited largely to experimentation with multiple regression or correlation. Econometrics was just emerging in universities, and many graduates were not familiar with how to use the technique to aid decision making. The emergence of the electronic computer allowed more widespread training, experimentation, and application of demand parameters and models to a variety of research problems.

Many senior business economists in the sixties were trained at a time when quantitative techniques and models were not widely used. Only more recent graduates were familiar with their development and use and few of them were in positions to significantly influence decisions. Equally important was the lack of adequate software and compilers to run computerized programs. These hurdles made early progress in model development and usage difficult.

EARLY APPLICATIONS

The author's own early experience with using demand models was in forecasting commodity prices. The attitude of seasoned managers who were unfamiliar with the technique was one of puzzlement. While receptive to the approach, they were cautious about its value and usefulness in meeting the needs of the organization. There were several developments that helped to successfully introduce the technique, however. First, the model fit the data reasonably well; second, the key variables included matched managers' own experiences; and third, the resulting forecasts were plausible. These factors helped immeasurably in explaining the results and gaining a degree of acceptability for the techniques.

Beginning practitioners are often enamored of the use of the quantitative technique rather than the resulting forecast and how it might be useful in decision making. With experience it becomes obvious that the technique is of less importance and of less interest than the result and how it is communicated. Experience also reveals there is no real substitute for knowledge of the industry, the marketplace, or the interactions of buyers and sellers. Modelers often substitute a technique of analysis for a lack of experience and understanding of the industry or market and what drives it.

But when both are combined, the results can be remarkably productive.

Industry analysts have found models most valuable when used to augment knowledge and experience rather than to replace it. A common mistake is to try to shortcut the decision process by using a model, as valid as it might be, to determine the correct course of action, and to make the decision. A model, when integrated with market knowledge, risk evaluation, and normal decision processes, can be an invaluable decision aid. But if used as a replacement for these, it often fails and is of little value in business.

The value of quantitative methods expanded beyond single equation models to large-scale simultaneous equation or sector models in the 1970s. This was the time when econometrics emerged to reach a much broader audience and to become a marketable, "commercial" product. The "what if" question in policy analysis and forecasting could be answered more easily with the more comprehensive model. The development of these complex models was a major advance in the application of quantitative techniques to complex industry behavior. They created widespread awareness and a general acceptance evolved.

Most of the industry, including some dairy firms, had some experience with these models. They were found to be helpful at times but disappointing in many respects. A major concern was in the forecast performance area. Many were poorly specified and yielded unreliable and unrealistic results that destroyed their credibility.

Many business economists either reviewed or subscribed to the major economic and agricultural models at one time or another and found them unwieldy and unreliable for the most part. However, improvements were made and one should not be overly critical of the advances and accomplishments of these modeling efforts. When properly understood and used judiciously, they were valuable for a number of business purposes.

Out of frustration and disappointment with the superficiality and performance of these models, many business economists turned to internally developed subsector models for their industries. These efforts were sometimes made in concert with the major econometric consulting firms. The level of detail and sophistication required for proper specification called for significant expenditures of staff, effort, and time. The final product, however, was a more sound, workable model that tracked the industry performance much better.

Modern Experience

Applications of econometric models and modeling in the business world are growing more common. After early experiences with large-scale models, industry has learned better their uses and limitations. These complex models have improved with industry input and involvement in their refinement. In many cases the models have reverted to less complex models with linkage and logic better understood and tracked. And, the theory on model formulation and estimation has improved.

Business applications are greater today with smaller single equation models being developed that are tailored for specific purposes (i.e., forecasting, elasticities, and variable significance). The larger multiequation simultaneous models have become most useful for two groups: policymakers and policy analysts and organizations serving a principal industry (e.g., banks, retailers, trade associations, suppliers servicing the dairy industry). Single equation models and sector models are more common within a particular industry such as the dairy industry. Even here their use is limited largely to national or multinational firms with larger staffs. Smaller, intermediate-sized firms often do not make use of demand models at all because of inexperienced staff and cost of development.

Large econometric models are predominantly used today for policy analysis and for indicating broad trends within an industry. Their usefulness in commodity market forecasting has diminished greatly within most business firms. Within the dairy industry little use is made of them except perhaps as a check against internal analysis and forecasting. Simpler and more specific, single equation demand and supply models are more common today in business. Academic institutions, government agencies, and consulting firms still form the predominant user group for large-scale models. Here the focus is more on long-range forecasts and policy analysis and evaluation.

Specific Business Uses

The evolution of business use from simple single equation models to large-scale multisector models and back to multiequation sector models (e.g., dairy industry) and finally even to simpler models for specific purposes represents a heightened awareness of specific needs of business. This development is even more understandable when one considers the orientation of the business economist, the evolution of model development, and trends in computer size and technology. The needs for business purposes are clearly different from those of other institutions and experience has focused attention on those approaches that address these specific needs.

Demand models are used for many different purposes in business, and attempts to enumerate or classify their uses run risks of omission. Nevertheless, the principal uses in business and examples of how they are used include the following:

- Forecasting business environments and performance for long-range plans
  - economic indicators
  - industry growth

- Commodity price forecasting and risk evaluation
  - sales and revenue forecasts
  - procurement strategies
  - market positions

- Market positioning and strategy
  - market plans and budgets
  - promotion and advertising decisions
  - inventory plans
  - investment decisions

- Dairy policy evaluation and formulation
  - production, consumption, price levels
  - surplus levels and costs
  - allocations of promotion dollars
  - support price evaluation

One common use of demand models in business is for projecting the environment in which business will operate. Most firms today develop annual business plans and long-range plans where the economic environment and the industry environment must be forecast. Models are useful in projecting economic growth rates, inflation levels, demand levels for key products, and commodity prices. Occasionally they are used to forecast a firm's own sales of a particular product. More often, however, they will be used to forecast industry sales and prices. From industry forecasts the firms project sales volumes and market shares based upon planned market actions.

Demand models are frequently used to forecast commodity market outcomes. In particular, market-determined prices for commodities, as opposed to government supported commodities, are fertile ground for demand modeling. The focus is to determine monthly, quarterly, or annual prices for the commodity. These forecasts along with technical market analysis and judgment become the basis for forward procurement decisions. In other words, if the price of a commodity has moved well below its forecasted equilibrium price due to seasonal or technical factors, a firm may decide to accumulate or book the raw material or product ahead. This practice is far more common in nondairy commodities than within the dairy industry. In part this is because of price volatility, industry practice, and hedging abilities in nondairy commodities.

Firms with commodity risk policies that gauge the risk associated with commodity ownership positions may use price forecasts for determining risk levels. For example, ownership of a commodity near or above its forecast price and/or its historical high, represents large inventory price risk that must be weighed against possible product shortage. Also, determining the demand response to a particular price level will aid in evaluating the ownership risk and the likely duration of a price extreme. This evaluation may be aided by a model and forecasts, but sound judgment and experience are required.

Major firms that market products beyond the primary commodity level usually develop annual marketing plans and long-range plans. These plans include sales forecasts, market strategies, pricing strategies, market segmentation decisions, and advertising and promotional budgets. To determine these factors marketers want to know industry growth possibilities by product, commodity price forecasts, and relevant developments and price forecasts of closely substitutable products. Often a firm will use demand models to arrive at these forecasts since it has the capacity to simultaneously incorporate the various factors affecting consumption of a product (i.e., price, income, prices of substitutes, trend).

Inventory decisions on basic high-demand products are driven by price expectations along with the firm's proclivity for risk and profit maximization. If price of raw material or finished product is forecast to rise strongly, additional inventory may be accumulated to ensure supply or to capitalize on market price appreciation. Again, demand models can alert management to the likelihood of these price movements. Because of their impact on profit, such decisions on price, inventory level, and associated risks are made with unusual care. Superior information or intelligence is at a premium. Either marketing-oriented or production-driven firms can capitalize on this information and pursue appropriate strategies.

Since the U.S. dairy industry operates under a price support program and federal market orders, policy options are constantly being evaluated within the industry and government. Price support and expenditure decisions are driven by forecasts of supply and demand for dairy products. These forecasts are derived directly from demand models and supply forecasts. Both the government and private industry utilize several different models to make these critical forecasts. The level of forecasted CCC (Commodity Credit Corporation) purchases has become a vital trigger level for price support changes since the Farm Security Act of 1985. A key determinant is expected consumption at existing price levels. Private industry also focuses upon expected demand levels to determine CCC's availability of surplus products for various government food and nutrition programs. These, of course, directly impact commercial sales, market strategy, and supply sourcing within industry.

One of the more obscure uses of demand parameters is in the allocation of price support changes between joint products: cheese and whey, butter and nonfat dry milk. From time to time the relative proportion of price decreases (increases) will shift between butterfat and nonfat solids. Recent changes from an equal weighting to a 100 percent butterfat, 0 percent nonfat dry milk allocation reflect the relatively more elastic demand for nonfat dry milk. Government and industry analysts have agreed that a change in the weighting of price changes would be beneficial, cost effective, and responsive to marketplace changes for the respective products. This is an excellent example of empirically derived demand parameters supporting industry and government judgment and intuition.

LIMITATIONS OF MODELS

Any evaluation of econometric models must recognize their multiple types, forms, and designs. For discussion purposes, if one were to focus only on single equation demand models and large-scale multiequation models, most business applications would be covered. Business use of such models is limited largely to only two functions: (1) forecasting and (2) policy analysis. Other valid research and business uses are not addressed.

Forecasting is one of the principal uses of econometric models although not the all-inclusive function that many in industry believe it to be. Most in industry who use econometric models employ them for forecasting purposes. Often they are used to forecast to the exclusion of other analytical devices (cycles, seasonals, graphics, industry structure, and so forth). And, as one might expect, the models are often misused and called upon to deliver far more than they are designed for.

SIMPLICITY VERSUS COMPLEXITY

Single equation models are extremely useful and workable in forecasting market prices, sales, market production, and markets that respond to normal market forces. They have the advantage of being simple, understandable, inexpensive to develop and operate, and flexible enough to be applied to many different time dimensions with only moderate skill levels required. Many analysts have found them useful in providing benchmark forecasts of milk production, consumption levels of milk, butter, cheese, and other dairy products. Their use in forecasting dairy product prices has been less effective than basic fundamental, subjective or naive analysis.

Commodity markets are much more complex than available data bases portray. Therefore, models often are not adequate by themselves in market decision making. They are valuable as aids but must be integrated with a number of analytical techniques, industry experience, and risk analysis to be effective. This seems to be even more important in *commodity* price forecasting than in other uses. One of the reasons may be because fundamentals of the marketplace fail to adequately capture the interaction of buyers and sellers and the timing aspects of this interaction.

The value of large-scale, multiequation models is primarily in describing in quantitative terms complex sets of interrelationships and linkages in major industries and markets. They are useful in creating an understanding of the impacts of changes in policy variables and estimating their quantitative impact. No other analytical tool possesses the ability to link sectors, markets, and control variables in an understandable fashion. Their usefulness in forecasting specific markets and industries, however, is less than in showing interrelationships and impacts of changes in policy variables. Many users, however, will use them for both purposes.

One of the ongoing problems with large-scale models has been a tendency for the focus to be more on technique and less on results,

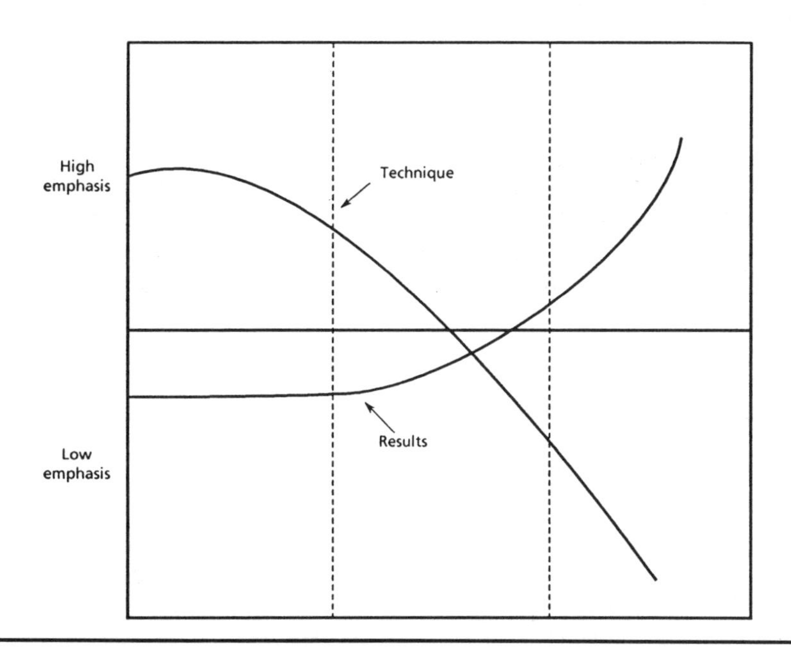

Figure 3.1. Evolution of econometric models

particularly in the early stages of their use (see Figure 3.1). As they become less of a novelty and more widely used, the emphasis has increasingly focused on results. In the mature stages, which we may now be entering, the focus intensifies on results, accuracy, and how to best use the model, and much less on the model per se. This evolutionary process is not widely different from many other processes in the business world.

Data Requirements

Large models are effectively used by institutions to understand and forecast sectors they are not intimately involved in but depend upon. Understanding the broader aspects of an industry's behavior and performance is certainly important and of great value to government, researchers, policy analysts, banks, supply industries, and others. However, for a firm in that particular industry much of the interrelationship, linkage, and impact is known from experience and judgment. These firms are less inclined to see the value of the model.

The limitations of large models are principally in their misuse and their large data requirements. They seldom incorporate the major external factors that often have the most impact on the final results. They also require considerable professional expertise and maintenance that often are not available to firm decision makers. Decentralized operations and small businesses are seldom adequately staffed to avail themselves of such costly and complicated analytical tools. These drawbacks will be partially overcome in time with the advent of smaller personal computers, but it remains doubtful that large-scale models will be used extensively at this level.

A large model's data requirements demand considerable time and effort for updating and revision. This time could be more profitably spent within a business experimenting and refining models and in actual applications. Usually, business is thinly staffed and can ill afford the necessary time or the cost. However, data base updates are usually available by outside vendors and can be procured. Nonetheless, the large model's value is diminished within business and often will be replaced by a simpler model with less maintenance overhead.

Integration with Management Systems

One of the most overlooked areas in model specification and development is input from the industry being modeled. Models are often developed from the academic view without adequate participation and input from knowledgeable industry analysts and decision makers. This is a serious omission and problem with many models, since they then may fail to capture the unique subtleties of an industry. Acceptance of a model and eventual use will evolve more easily if this collaboration occurs and is nurtured.

The true value of a model can often be determined by how effectively it becomes integrated into management decision-making systems. When it becomes a regular ongoing part of internal business systems, one can be assured it is reaching key decision makers. Usually this involves active involvement and participation by the analyst or economist in key operational planning. The complexity of the modeling technique often requires this involvement in the early stages of use. Later, as informal training occurs and familiarity develops, the models can be turned over to less specialized personnel.

Those involved in the development of models are advised to consult with and draw upon the expertise residing in industry. It is not a bad reflection on the model developer's talent to seek better understanding of the internal workings and nuances of complex industries. There is little to be lost and much to be gained with greater consultation—principally a

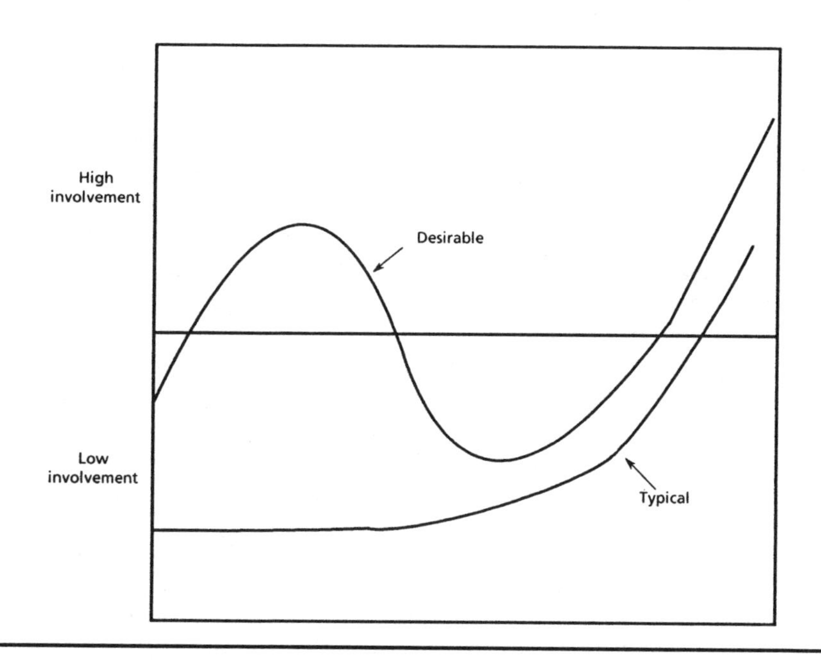

Figure 3.2. Degree of industry involvement in model development

better and more usable product. Figure 3.2 depicts two levels of involvement on the part of industry personnel when the model is developed outside the firm or industry: the more typical involvement and a more desirable level of involvement. Even when all development occurs within a firm, the model developer is advised to involve key operational personnel extensively. An exception is when operational management has strong confidence in the staff member developing the model and routinely involves that person in key communication loops.

IMPROVEMENTS

Developers of large complex models need to focus on several areas to improve overall use and effectiveness of these models in business. The principal areas for improvement include

- Data bases
- Representation of unique industry characteristics
- Performance
- User understanding
- Cost of development and operation
- Inclusion of key external factors
- Timeliness

Industry is particularly skeptical if models fail to capture key turning points and miss directionally the impacts of key developments. This has been the case with many large-scale models, and the associated loss of credibility has been damaging to greater model use. Similarly, large adjustment factors or "tweaking the model" on a regular basis tends to destroy confidence in its use. This does not suggest that models are of no value, but it does reduce reliance on them for decision purposes.

Model vendors need to perform more postanalysis to determine their effectiveness under different conditions and environments. These evaluations, if shared with users, will help them in decision making and will lend additional credibility to the demand models. No one expects perfection with a model, but some reasonable error range should be provided and should prove acceptable to most users. Many of the improvements in models and their use will depend upon the degree of customer focus by developers. The identification of the customers' needs and directing effort toward the achievement of those particular needs will accelerate the acceptance and wider usage of demand models in decision making.

AREAS OF NEW RESEARCH

There have been numerous encouraging efforts in the last several years in the direction of needed research and development. The work of Womack and Johnson at the University of Missouri and Iowa State, respectively, on government policy alternatives, and of Goddard at Guelph and of Kinnucan at Georgia and Auburn University on generic advertising of dairy products, and Huang and Haidacher at USDA on food system demand and elasticities, are examples. These recent works and others of a similar nature are valuable to decision makers and are worth pursuing.

There are several areas of needed research where econometric models can focus and serve practitioners, policymakers, and business to a greater degree than before. If pursued diligently and skillfully major advances can be made in the use and understanding of models. Priority areas for research and development are

- *Government farm and food program analysis.* Further refinement of models and examination of alternative programs or delivery systems would be beneficial. More attention and evaluation of cost/benefit program objectives and outside shock factors would prove enlightening.
- *Export policy and international trade.* The allocation of funds for various competing programs, the payback, along with the emphasis of these programs, would aid policymakers and industry greatly. Their performance under different environments (economic and political) would add to our understanding.
- *Commodity promotion and advertising.* Generic versus brand advertising, promotion versus pure advertising, and levels of expenditure under each method need careful analysis. The optimization-allocation question of limited funds for these programs deserves attention also.
- *Product marketing.* The evaluation of pricing strategies for mature products versus growth products and new derivative products would aid the industry. The question of the age-old promotion/advertising mix would be a fruitful research area and of value to business.

These research areas were selected because of their high impact upon farming, government, agribusiness, and the inadequate analytical efforts currently underway. Few would question the impact of farm and food distribution programs on dairy farming, agriculture, or rural communities in the next decade. We need to ask if we have utilized our analytical tools to their fullest to aid policymakers. Forecasts of impacts at different program levels would be useful to policymakers, for example, for the Conservation Reserve Program, land removal up to 50 percent of an area (county or region), and its impact on rural communities.

International trade and export policies pursued by various countries around the world impact the results of domestic programs and the welfare of many related sectors. The modeling efforts in this area are limited, and it is doubtful that we have addressed these questions adequately at this point. Can demand models play a greater role in these evaluations?

The funding of dairy promotion programs in the United States to relieve product surpluses and expand consumer demand offers an excellent opportunity to apply analytical tools to a real business problem and to aid decision makers tremendously. We need more conclusive research to indicate the best allocation of advertising and promotion dollars among (1) products, (2) geographic markets, (3) season and frequency of promotion, and (4) forms of advertising/promotion. Overall, the effectiveness of these expenditures needs more definitive examination. Unfortunately, much of this research must now be done simultaneously with large expenditures of funds being made without adequate guiding research. This tends to dilute the effectiveness of research efforts and very likely the return on invested dollars.

Business should be urging more product-specific modeling to be used by marketing personnel in agribusiness and the food industry. For example, could demand models accurately measure the effectiveness of promotional dollars, advertising dollars, and other marketing variables on sales of mature products, products in growth markets, and premium versus discount products? This refinement would likely cause marketing decision makers to utilize econometric demand models as decision tools much more than to date. Some firms have already utilized models for this purpose, but it is not very common in business.

SUMMARY

Econometric demand models have advanced the cause of economic analysis during the last 20 years. Their acceptance and usefulness in decision making has grown but appear to lag their promise and potential considerably, especially in business. The dairy industry tends to make greater use of demand parameters because of government policy options and key trigger levels more than most other industries. However, their usage in operational analysis and marketing appears not to be extensive at this point. Large national and multinational firms are believed to make much greater use of models than smaller to intermediate sized firms.

Skilled use of models undoubtedly will add to judgment, experience, and the acumen of decision makers in the business world. The needed improvements in modeling and in coordination with decision makers are somewhat overlooked, but will grow as the process of integrating demand models matures and new decision makers become more familiar with the techniques. Model developers clearly need to be more customer oriented if they are to achieve successful integration into management decision systems.

New areas of research focus include the orientation of researchers and models to a number of major policy arenas including farm policy, trade policy, commodity promotional programs, and branded product marketing. Additional research and application in these areas will improve the usefulness and acceptance of demand models in the business world.

REFERENCES

Goddard, E. W. and A. Tielu. 1988. "Assessing the Effectiveness of Fluid Milk Advertising in Ontario." *Canadian Journal of Agricultural Economics.* 36(2):261–278.

Goddard, E. W. and A. Amuah. 1989. "The Demand for Canadian Fats and Oils: A Case Study of Advertising Effectiveness." *American Journal of Agricultural Economics.* 71(3):741–749.

Huang, K. S. and R. C. Haidacher. 1987. *An Assessment of Price and Income Effects on Changes in Dairy Consumption.* Paper presented at Dairy Products Demand Symposium, Atlanta, Ga., October.

Huang, Kuo S. 1985. *U.S. Demand for Food: A Complete System of Price and Income Effects.* TB-1714, U.S. Department of Agriculture Economics Research Service, December.

Johnson S. R. and A. W. Womack, 1987. *FAPRI Ten-Year International Outlook.* FAPRI Staff Report 4-87. Iowa State University and University of Missouri-Columbia, March.

Kinnucan, H. W. 1982. *Demographic Versus Media Advertising Effects on Milk Demand: The Case of the New York City Market.* Cornell Agricultural Economics Staff Paper No. 82-5. Ithaca, N.Y.

Kinnucan, H. W. 1986. "Importance of Functional Form in Optimizing Advertising Expenditures for Fluid Milk." In *Workshop on Demand Analysis and Policy Evaluation.* Edited by D. Peter Stonehouse, Bulletin No. 197/1986. Brussels, Belgium: International Dairy Federation.

Kinnucan, H. W. and O. D. Forker. 1986. "Seasonality in the Consumer Response to Milk Advertising: Implications for Milk Promotion Policy." *American Journal of Agricultural Ecnomics.* 68(3):562–571.

CHAPTER 4

Empirical Demand Systems

Tesfaye Teklu, Zuhair A. Hassan, S. R. Johnson, and D. Peter Stonehouse

INTRODUCTION

MARKET DEMAND systems methods are receiving increased application in the study of consumption patterns for dairy products. In part, this is a result of developments in the theory and method of demand systems estimation. On a more pragmatic level, the application of these methods has been stimulated by the importance of substitution effects for price and income changes and concerns about the characteristics of participants in dairy product markets. In the former case, demand systems provide hypotheses which can be tested on substitution and specifications that can be rationalized using more clearly understood assumptions about the underlying preferences of consumers. For the latter, more systematic scaling and translating methods are available for studying the impacts of changes in sociodemographic characteristics on consumer demand parameters. In both cases, the result has been improved robustness and reliability of estimated demand parameters for dairy products.

The purpose of this chapter is to review developments for estimating market demand functions that include sociodemographic characteristics. Review of the theory is somewhat technical, although required to demonstrate the robustness of the results currently available for estimating demand systems at the market level. The market demand systems being applied require restrictive assumptions. An alternative is to locally approximate the demand systems. For the inclusion of sociodemographic effects, available results are more conclusive for guidelines in specification. Scaling and translating methods can be incorporated in demand systems estimation for tests of the theory and for policy analysis.

The discussion proceeds from a brief review of the consumer demand theory. This includes selected modern demand systems, rationalized on the basis of correspondences between individual and market level responses. The types of demand systems reviewed are commonly applied in food and agricultural policy analysis. Benefits of scaling and translating for incorporating sociodemographic affects are then assessed. More recent attempts to address the correspondences between demand at individual and market levels are next investigated. The final section provides observations on the advantages of more directly using the theory in applied analysis and promising directions for future research.

BASIC CONSUMER DEMAND THEORY

The review of basic consumer demand theory is organized into three parts. Part one features the direct approach to the utility maximization problem. The dual version of the consumer allocation problem is presented in part two. The more common of the empirical demand models are surveyed in part three.

Utility Maximization

The consumer allocation problem can be formulated in a utility maximization framework (for example, see Deaton and Muellbauer 1980a; Phlips 1983; and Johnson, Hassan, and Green 1984). The problem is to find an optimal vector $Q^*(P,Y)$ at which the consumer attains maximum utility $U^*(Q)$, subject to the linear budget constraint, i.e.,

$$\begin{aligned} &\text{Max } U = U(Q) \\ &\text{St. } P'Q = Y. \end{aligned} \tag{4.1}$$

This maximization, under plausible conditions on the utility function, yields the system

$$\begin{aligned} Q^* &= \Phi(P, Y) \text{ and} \\ \lambda^* &= \lambda(P, Y), \end{aligned} \tag{4.2}$$

where Q^* is an n-element vector of optimal quantities expressed as functions of the prices (P) of goods and income (Y); λ is the Lagrangian multiplier interpreted as the marginal utility of income.

Applications of this model of consumer behavior have primarily involved the estimation of the parameters for the system $Q^* = \Phi(P, Y)$. For n commodities, there are a total of $n^2 + n$ elasticities to be estimated, n^2

price elasticities and n income elasticities. The number of unknown elasticities can be reduced by imposing restrictions from the consumer optimization problem and properties of the utility function. As well, by assuming special functional forms for the utility and/or demand functions—limiting the generality of the choice problem—the number of elasticities to be estimated can be further reduced.

The restrictions on the demand parameters derive from the properties of the utility function and the linear budget constraint. These can be illustrated using the solution to the fundamental matrix equation of consumer demand theory (Barten 1964) which yields

$$\frac{\partial Q}{\partial P} = \lambda U^{-1} - \lambda(\partial\lambda/\partial Y)^{-1}\left(\frac{\partial Q}{\partial Y}\right)\left(\frac{\partial Q}{\partial Y}\right)' - \left(\frac{\partial Q}{\partial Y}\right)Q', \tag{4.3}$$

$$\frac{\partial Q}{\partial Y} = \left(\frac{\partial \lambda}{\partial Y}\right)U^{-1}P, \tag{4.4}$$

$$\frac{\partial \lambda}{\partial P} = -\lambda\left(\frac{\partial Q}{\partial Y}\right) - \left(\frac{\partial \lambda}{\partial Y}\right)Q, \text{ and} \tag{4.5}$$

$$\frac{\partial \lambda}{\partial Y} = (P'U^{-1}P)^{-1}, \tag{4.6}$$

where $\frac{\partial Q}{\partial P}$, $\frac{\partial Q}{\partial Y}$, $\frac{\partial \lambda}{\partial P}$, and $\frac{\partial \lambda}{\partial Y}$ are the derivatives of the demand equation (4.2) with respect to the prices, P_i, and income, Y, and U^{-1} is the inverse of the Hessian matrix, U. Because of the continuity and differentiability assumptions, U is symmetric. A strict quasi-concavity assumption for the utility function implies that the Hessian is negative definite. The symmetry and the negative definite properties of the Hessian matrix imply that the substitution matrix,

$$K = \lambda U^{-1} - \lambda\left(\frac{\partial \lambda}{\partial Y}\right)^{-1}\left(\frac{\partial Q}{\partial Y}\right)\left(\frac{\partial Q}{\partial Y}\right)', \tag{4.7}$$

is symmetric, with negative diagonal elements. The latter assures that the income compensated own-price elasticities are negative.

Expressing equations (4.3) through (4.6) in elasticities, the demand functions can be shown to satisfy three restrictions.

Engel aggregation:

$$\sum_i w_i \eta_{iY} = 1, \tag{4.8}$$

where $w_i = P_i Q_i / Y$ is the average budget share for the ith commodity. Budget exhaustion for given income, $P'Q = Y$, implies that the sum of the weighted income elasticities adds to unity. Thus, only $n - 1$ of the income elasticities are independent.

Homogeneity:

$$\sum_j \epsilon_{ij} + \eta_{iy} = 0 \qquad \text{for all } i, \tag{4.9}$$

where the demand functions are homogenous of degree zero in prices and income. That is, an equal proportional change in P and Y will leave commodity demands unaffected. For each demand function, there is therefore one redundant elasticity, and for n functions there are n redundant elasticities.

Slutsky Symmetry:

$$\frac{\partial Q}{\partial P} = K - \left(\frac{\partial Q}{\partial Y}\right) Q', \tag{4.10}$$

where K is the substitution effect and $\left(\frac{\partial Q}{\partial Y}\right) Q'$ is the income effect of a price change. From equation (4.10) the substitution matrix can be expressed as,

$$\epsilon_{ij} = \frac{w_j}{w_i} \epsilon_{ji} - w_j(\eta_{iY} - \eta_{jY}). \tag{4.11}$$

Equation (4.11) reduces the number of independent elasticities to $1/2(n^2 - n)$.

These three restrictions, Engel aggregation (eq. [4.8]), homogeneity (eq. [4.9]), and symmetry (eq. [4.10]), reduce the number of parameters to be estimated for the demand system to $1/2(n^2 + n - 2)$.

DUALITY

An alternative approach for obtaining the Marshallian demand system is to start from an indirect utility function,

$$V(P, Y) = \text{Max}\ [U(Q) : P'Q = Y], \tag{4.12}$$

where $V(P, Y)$ is the maximum attainable utility level for a given vector of prices and income. Applying Roy's identity to equation (4.12) produces

$$Q_i^*(P,Y) = V_{P_i}/V_Y \ \text{ for all } i \tag{4.13}$$

where $V_{P_i} = \dfrac{\partial V}{\partial Y}$ and $V_Y = \dfrac{\partial V}{\partial Y}$ are partial derivatives of the indirect utility function with respect to price of commodity i, P_i, and income Y, respectively.

The demand functions, equations (4.13) and (4.2), are identical. The choice of method for obtaining them depends on the observability of the required variables (Varian 1978).

Alternatively, suppose there is a target level of utility, U^*. The minimum income level required to reach U^* is

$$E(P, U^*) = \text{Min}\ [PQ : U(Q) = U^*], \tag{4.14}$$

where $E(P, U^*)$, the expenditure or cost function, is expressed for fixed prices and the given utility level. Shephard's lemma can be applied to the cost function (eq. [4.14]) to obtain

$$\frac{\partial E(P,U^*)}{\partial P_i} = Q_i^c(P,U), \tag{4.15}$$

where $Q_i^c(P, U)$ is a Hicksian demand function (utility constant instead of income constant), expressed in prices and U^*. When U^* is the level that represents the maximum utility in the primal problem, that is, $V(P, Y) = U^*$, both the indirect utility and cost functions yield the same demand functions. That is, the Marshallian and Hicksian demand functions are equal: $Q_i^c(P, Y) = Q_i(P, E)$. Moreover, the cost function, $E(P, U^*)$, can be obtained by inverting the indirect utility function, $V(P, Y)$. This is possible since $V(P, Y)$ is nondecreasing in Y. Similarly, the indirect utility function can be derived from the cost function by setting $E(P, U^*) = Y$ and solving for $V = U^*(P, Y)$. If we redefine the problem in equation (4.14) as

$$D(Q, U) = \text{Min} \; [PQ : V(P, Y) = U^*] \,, \qquad (4.16)$$

then $D(Q, U)$ represents the distance function, expressing the minimum cost of achieving a utility level U^* at a given vector Q^*. The distance function yields the vector of prices that will give the amount (proportion) by which Q^* must be divided to achieve U^*.

EMPIRICAL DEMAND SYSTEMS

Empirical applications of demand systems are now common place. Various demand systems have been used, defined by special forms implied for the utility function (Frisch 1959; Phlips 1983; and Johnson, Hassan, and Green 1984). One objective of these special forms is to insure that the results for the individual consumer allocation problem apply in the available aggregate or market data. These special forms and the conditions for aggregation incorporate varying forms of separability or near separability (Phlips 1983). Selected demand systems are reviewed to illustrate the advantages of the special forms and the associated restrictions.

Linear Expenditure System (LES). The LES demand function for commodity i is

$$Q_i = \gamma_i + \beta_i/P_i \; (Y - \sum_j P_j \gamma_j), \qquad (4.17)$$

where γ_i is interpreted as the committed quantity of commodity i and $Y - \sum_i P_j \gamma_j$ as supernumerary income, which the consumer allocates in fixed proportions, β_i/P_i. The demand system can be derived from a translated version of the Bergson family of utility functions (Pollak 1971), attributed to Stone and Gary,

$$U(Q) = \sum \beta_i \ln(Q_i - \gamma i), \qquad (4.18)$$

where $\sum \beta_i = 1$, $0 < \beta_i < 1$, and $(Q_i - \gamma_i) > 0$. With these restrictions on the utility function, the demand system satisfies adding up ($\sum \beta_i = 1$) and symmetry ($0 < \beta_i < 1$ and $Q_i > \gamma i$).

The LES incorporates the restrictions implied by an additive utility structure. The condition $0 < \beta_i < 1$ implies income elasticities are positive. The fact that the cross-substitution terms are positive implies that all pairs of goods are net substitutes. Also, for the LES specified for a large number of commodities, the price elasticities are approximately proportional to expenditure elasticities (Deaton and Muellbauer 1980a). Despite these limitations, the LES has produced plausible results, especially when goods are broadly grouped and price variations within these groups are limited (Phlips 1983).

Indirect Addilog Demand Model. The addilog demand system is derived from the additive, indirect utility function

$$V(P,Y) = \Sigma\alpha_i(Y/P_i)^{b_i}, \tag{4.19}$$

with parameter restrictions $\alpha_i < 0$, $\Sigma\alpha_i = -1$, and $-1 < b_i < 0$ (Houthakker 1960). The corresponding demand function for the ith commodity, in log form, is

$$\ln Q_i = \ln\alpha_i b_i + (1 + b_i)\ln(Y/P_i) - \ln\Sigma\alpha_j b_j(Y/P_j)^{b_j}, \quad i = 1, 2, \ldots, m. \tag{4.20}$$

The demand equation (4.20) is easily shown to satisfy the general restrictions from consumer demand theory (Johnson, Hassan, and Green 1984).

Almost Ideal Demand System (AIDS). Using the dual formulation of the consumer allocation problem, Deaton and Muellbauer (1980b) specified the cost function

$$\ln C = \alpha_0 + \Sigma_j\alpha_j\ln P_j + 1/2\,\Sigma\Sigma\gamma_{jk}\ln P_j\ln P_k + U\beta_0\Pi P_j^{\beta_j}, \tag{4.21}$$

where for the function to be linearly homogeneous in prices, the parameters must satisfy $\Sigma_i\alpha_i = 1$, $\Sigma_j\gamma_{jk} = \Sigma_k\gamma_{kj} = \Sigma_j\beta_j = 0$. Applying Shephard's lemma to equation (4.21) yields the Hicksian demand function for commodity i (in share form),

$$w_i = \alpha_i + \Sigma\gamma_{ij}\ln P_j + \beta_i U\beta_0\Pi P_k^{\beta_k}, \tag{4.22}$$

where $\gamma_{ij} = 1/2(\gamma_{ij} + \gamma_{ji})$ is required to satisfy the symmetry condition. Using the duality relation $Y = C(P, U)$, the indirect utility function, $U(P, Y)$, corresponding to equation (4.21) can be expressed as

$$U = \ln Y - (\alpha_0 + \Sigma_j\alpha_j\ln P_j + 1/2\,\Sigma\Sigma\gamma_{jk}\ln P_j\ln P_k)/\beta_0\Pi P_j^{\beta_j}. \tag{4.23}$$

Substituting equation (4.23) into equation (4.22) yields the Marshallian demand function for commodity i in share form,

$$w_i = \alpha_i + \Sigma_j \gamma_{ij} \ln P_j + \beta_i \ln \bar{Y}, \tag{4.24}$$

where $\bar{Y} = Y/P^0$ is nominal income deflated by a price index, P^0. The price index P^o is

$$\ln P^0 = \alpha_0 + \Sigma\, \alpha_j \ln P_j + 1/2\, \Sigma\Sigma \gamma_{jk} \ln P_j \ln P_k. \tag{4.25}$$

Equations for the full set of commodities consistent with equation (4.24) exhibit the restrictions from the standard consumer demand model if

$$\Sigma_i \alpha_i = 1, \quad \Sigma_i \gamma_{ij} = 0, \quad \Sigma_i \beta_i = 0; \tag{4.26}$$

$$\Sigma_j \gamma_{ij} = 0; \text{ and} \tag{4.27}$$

$$\gamma_{ij} = \gamma_{ji}, \tag{4.28}$$

where equations (4.26), (4.27), and (4.28) ensure the Engel aggregation, homogeneity, and the Slutsky symmetry conditions, respectively. The adding-up and homogeneity restrictions simply repeat the restrictions imposed on the parameters of the cost function. These restrictions can be applied to equations (4.24) and (4.25) to test the consistency of the demand system with demand theory (Brown, Green, and Johnson 1986).

For estimation, the price index P^0 can be approximated using Stone's index, $\ln P^0 = \Sigma w_k \ln P_k$. The one application available for evaluating this approximation suggests that it is reasonably accurate (Brown, Green, and Johnson 1986). The advantage of the approximation is that the demand system is linear in the structural parameters.

Indirect Translog Model (ITL). Instead of starting with a specific indirect utility function, Christensen, Jorgenson, and Lau (1975) approximate the true indirect function with a second-order Taylor series expansion. The approximation, $\psi(\hat{P})$, is in logarithms of income normalized prices, $\hat{P}_j = P_j/Y$,

$$\psi(\hat{P}) = \alpha_0 + \Sigma \alpha_i \ln \hat{P}_i - 1/2 \Sigma\Sigma b_{ij} \ln \hat{P}_i \ln \hat{P}_j, \tag{4.29}$$

where $\Sigma \alpha_i = -1$, $b_{ij} = b_{ji}\ \forall_i$ and j, and $\Sigma_j b_{ij} = 0\ \forall_i$. Using Roy's identity, the demand function for commodity i in share form is

$$w_i = \alpha_i + \Sigma_j b_{ij} \ln \hat{P}_j / \Sigma \alpha_k + \Sigma\Sigma b_{kj} \ln \hat{P}_j. \tag{4.30}$$

Writing equation (4.30) in quantity terms yields

$$Q_i = P_i^{-1}(\alpha_i + \Sigma b_{ij}\ln P_j)/\Sigma\alpha_k + \Sigma\Sigma b_{kj}\ln P_j. \qquad (4.31)$$

The demand system can be estimated subject to the symmetry restrictions ($b_{ij} = b_{ji}$) and an equality restriction $\Sigma_j b_{kj} = 0$ for all k in all the demand equations. The full demand system requires estimation of $1/2(n^2 + 3n - 2)$ parameters. Compared to the other demand systems reviewed, the ITL requires more sample information, since the number of parameters to be estimated is comparatively large.

If $b_{ij} = 0$ for all i and j, the indirect utility function equation (4.29) reduces to a simple Cobb-Douglas form. These utility functions (both the direct and indirect) are self-duals, that is, they represent the same preferences. This is the case with the linear logarithmic system popularized by Lau and Mitchell (1975). The imposition of this restrictive homothetic structure reduces the number of parameters to be estimated, but the demand system becomes much less flexible and behaviorally less plausible for food consumption analysis.

Rotterdam Demand Model (RDM). Unlike the above demand systems, the RDM started from a specific algebraic demand system, then the restrictions were imposed to make it consistent with the theory of consumer demand (Theil 1965; Barten 1969). The relative price version of this system has been subsequently shown to be derived from Stone's (1954) logarithmic demand function (Barten 1969):

$$\ln Q_i = \alpha_i + \eta_i\ln(Y) + \Sigma_j\epsilon_{ij}\ln P_j. \qquad (4.32)$$

Writing equation (4.32) in differentials yields

$$d\ln Q_i = \eta_i d\ln(Y) + \Sigma_j\epsilon_{ij}\ln P_j. \qquad (4.33)$$

After algebraic transformations, equation (4.33) can be written in estimable form (Thiel 1965)

$$w_i d\ln Q_i = \mu_i d\ln\bar{Y} + \Sigma_j b_{ij} d\ln\bar{P}_j \qquad (4.34)$$

where $\mu_i = P_i \dfrac{\partial Q_i}{\partial Y}$ is the marginal budget share, $\bar{Y} = d\ln Y - w_k d\ln P_k$ is a measure of real income, $b_{ij} = \lambda U_{ij}P_jP_i/Y$ is the coefficient of the relative price j, and $\bar{P}_j$ is the deflated price of the jth commodity. The demand equation (4.34) satisfies the restrictions for the consumer allocation problem; adding up ($\Sigma\mu_i = 1$), symmetry ($b_{ij} = b_{ji}$ for all$_{i,j}$), and homogeneity ($\Sigma_i b_j = \phi_i \mu_i$).

The parameters of the RDM can be significantly reduced if additivity restrictions are further imposed (Johnson, Hassan, and Green 1984). Then, only $n + 1$ parameters (μ_s and ϕ) are required for a complete set of demand elasticities.

The assumption of the constancy of the coefficients (μ_i, b_{ij}) in equation (4.34) implies a specific structure of the underlying utility function. It has been shown (Goldberger 1969; Yoshihara 1969) that the Rotterdam demand system can be derived from the Cobb-Douglas utility function,

$$U = \Sigma_i \beta_i \ln Q_i. \tag{4.35}$$

The demand system derived from equation (4.35) implies that income elasticities are all equal to unity, all own-price elasticities are equal to -1, and all cross-price elasticities are equal to 0. These results illustrate the restrictiveness of the RDM for empirical work.

A Local Box-Cox Approximation. The Box-Cox (1964) and other functional forms provide a basis for developing "local" approximations of demand systems. The Box-Cox form utilizes a transformation of variables defined by

$$C_i(\lambda) = Z_i(\lambda)\beta, \tag{4.36}$$

where $C_i(\lambda) = C_i^{\lambda} - 1/\lambda$ and $Z_i(\lambda) = Z_i^{\lambda} - 1/\lambda$, and λ is the transformation variable. The function can be specified to include a value, λ, for each variable. If one postulates that $C_i(\lambda)$ represents a demand function for a particular commodity i, equation (4.36) can be viewed as defining a general Box-Cox demand system (Johnson et al. 1985). The original demand system, equation (4.2), is replaced by

$$Q_i(\lambda) = \phi_i[P_i(\lambda), Y(\lambda)]; \; i = 1, 2, \ldots, n. \tag{4.37}$$

The corresponding restrictions from the demand theory are derived by Johnson et al. (1985), who note that the restrictions hold only locally at specified prices, income, and budget proportions. Associated expressions can be used to combine, in a mixed estimation context, prior and sample information to estimate local approximations to demand systems.

The most commonly used flexible functional forms also are special cases, if equation (4.36) represents a generalized Box-Cox cost function (Berndt and Kahlad 1979). For example, the function expressed as a quadratic in Box-Cox transformations is a translog as $\lambda \rightarrow 0$, generalized Leontief if $\lambda = 1$, generalized square root if $\lambda = 0.5$, and quadratic if $\lambda = 2$. The implication is that tests for values of λ can be used to select from an extended family of local flexible approximations.

INCORPORATING DEMOGRAPHIC VARIABLES

The utility function, equation (4.1), is specified at the household level and implicitly assumes that households with different socioeconomic characteristics have similar preference structures. This assumption likely has, however, limited validity in situations where socioeconomic characteristics (family size, age-sex composition, location, etc.) influence consumption behavior of the household unit. A more plausible approach is to respecify the utility function conditioned by these variables, that is, $U = U(Q/\eta)$ where $\eta = (\eta_1, \eta_2, \ldots, \eta_r)$ is a vector of demographic characteristics.

Scaling for household composition. A simple approach is to express quantities demanded and income, equation (4.2), in per capita terms. This specification, however, fails to incorporate variations for example, due to the age-sex composition of individuals in households. The classical approach to this problem is Engel's (1895) work which reflected differences in household composition in income-consumption relationships. The quantities and income in the demand functions are normalized in "adult equivalents,"

$$Q_i/m_0 = \bar{Q}_i(Y/m_0), \tag{4.38}$$

where m_0 is the adult equivalent index and a function of characteristics of household members.

Prais and Houthakker (1955) provided a commodity-specific generalization of Engel's specification with adult equivalent scales, m_i's, defined separately for each commodity,

$$Q_i/m_i = \bar{Q}_i(Y/m_0), \tag{4.39}$$

where m_i is a commodity-specific scale and m_0 is a general or income scale. That is, m_0 is a weighted average of the individual commodity scales, $m_0 = m_0(m_1, m_2, \ldots, m_n, Y)$. Note that these commodity-specific scales incorporate no relative price effects. The Engel and Prais-Houthakker specifications are, hence, consistent with demand theory where prices are held constant by experimental control, as in cross-section data.

Scaling generalized to prices. Barten (1964) presented a generalization of the Prais-Houthakker work in a modified utility framework. The resulting modified demand function is

$$Q_i = m_i\bar{Q}_i(P_1m_1, P_2m_2, \ldots, P_nm_n, Y) \text{ for all } i, \tag{4.40}$$

or, in scaled quantities and prices,

$$\hat{Q}_i = \hat{Q}_i(\hat{P}_1, \hat{P}_2, \ldots, \hat{P}_n, Y) \text{ for all } i, \tag{4.41}$$

where $\hat{Q}_i = Q_i/m_i$ is a "normalized" quantity of commodity i, m_i is a commodity-specific adult-equivalent scale, and $\hat{P}_i$ is a normalized price. Changes in family composition therefore modify relative prices and, consequently, substitution among goods.

The Barten approach has been defined by Pollak and Wales (1969, 1978, 1981) as a scaling technique for incorporating household-specific demographic variables in demand systems. The procedure involves first postulating the scaling parameters, $m_1, \ldots, m_n$, which depend only on demographic variables. That is,

$$m_i = \bar{m}_i(\eta) \text{ for all } i, \tag{4.42}$$

where the modified demand function is as specified in equation (4.41). Pollak and Wales (1981) reviewed and/or developed four other procedures for reflecting nonhomogeneous household effects: demographic translation procedure, the Gorman procedure, the reverse Gorman procedure, and the modified Prais-Houthakker procedure.

Translation. The translation procedure facilitates the introduction of parameters, $d_1, \ldots, d_n$, linked to the demographic variables through the functional form

$$d_i = d_i(\eta) \text{ for all } i. \tag{4.43}$$

The modified demand system replacing equation (4.2) is

$$Q_i = d_i + \bar{Q}_i(P, Y - \Sigma P_k d_k), \tag{4.44}$$

where only the d's depend on the demographic variables. These translating variables are interpreted to represent characteristics of all household members as opposed to scales, for example, race, region, location. The d's can be viewed as parameters reflecting "subsistence" or "necessary" consumption levels. Alternatively, the demand function in equation (4.44) can be viewed as being generated in a two-stage budgeting process. First, the household allocates part of its total expenditure to a vector of necessary quantities, $d_1, d_2, \ldots, d_n$. Then, in the second stage, it allocates the balance, $Y - \Sigma P_k d_k$, among the various commodities. This interpretation is similar to that advanced for the Linear Expenditure System (LES).

The Gorman Procedure. This procedure can be viewed as equivalent to first scaling and then translating the original demand function,

$Q_i = Q(P, Y)$. The modified demand system becomes

$$Q_i = d_i + m_i\bar{Q}_i(P_1m_1, P_2m_2, \ldots, P_nm_n, Y - \Sigma P_kd_k), \qquad (4.45)$$

where the d's and m's are parameters postulated to depend on sociodemographic variables. If the order is reversed, that is, the demand function is first translated and then scaled, the resulting procedure (reverse Gorman) yields a system

$$Q_i = m_i[d_i + \bar{Q}_i(P_1m_1, P_2m_2, \ldots, P_nm_n, Y - \Sigma P_kd_k)]. \qquad (4.46)$$

The Modified Prais-Houthakker. This procedure adjusts the original demand system (eq. [2]) as

$$Q_i = m_i\bar{Q}_i(P, Y/m_0), \qquad (4.47)$$

where the commodity-specific scales, m's, depend on the demographic variables, $m_i = \bar{m}(\eta)$. The income or composite scale, m_0, is defined through the budget constraint,

$$\Sigma P_im_i\bar{Q}_i(P, Y/m_0) = Y \qquad (4.48)$$

and a function of all prices, income, and demographic variables, $m_0 = \bar{m}_0(P, Y, \eta)$.

HOUSEHOLD PRODUCTION

The household production theory, primarily an outgrowth of the traditional consumer theory, has two major themes. First, the classical consumer demand theory assumes that a consumer derives satisfaction from the consumption of market goods. The "new" approach hypothesizes that the consumer derives satisfaction from consumption of characteristics, derived in part from market goods. These characteristics are produced in the household from the combination of purchased market goods, labor, and human and physical capital. Second, the explanatory variables in the classical demand functions are market prices, income, and "taste." The classical theory is incapable of either explaining or predicting effects of selected proxies for tastes on consumption behavior. The approaches to alter demand models demographically can be interpreted as, for example, attempts to reflect these characteristics, enhancing the explanatory power of demand theory.The household production approach rationalizes the incorporation of production-related parameters in demand systems.

The household production theory is essentially a combination of theory of the consumer and of the firm. The households, as producers, decide the mix and level of production of the home-produced consumption characteristics. Simultaneously, the household decides the amounts of the market-purchased goods, labor, and other factors to allocate to the production of these characteristics or final commodities. Efficiency of allocation of resources is guaranteed if the household minimizes the cost of producing the desired level of final commodities. As a consumer, the household allocates the income derived from the production activity to the consumption of the "commodities" from which utility is derived.

There are two variants of the household production theory: the commodity characteristics (Lancaster 1966) and the production function approaches (Becker 1965; Muth 1966; and Michael and Becker 1973). According to the characteristics approach, the final commodities represent objective features of market goods. Consumers buy commodities, not for the commodities per se, but for their characteristics, the primary objects of choice. The decision to allocate income, for example, to food purchases reflects a conscious and rational effort of consumers to purchase a desired mix of characteristics. Limitations of the Lancaster model are the assumption of linear consumption technology (Lucas 1975), the assumption that the utility function depends upon the level of characteristics and not on their distribution among commodities, and the nonnegativity of the marginal utility of characteristics (Hendler 1975).

Like the Lancaster model, the production approach assumes that consumers derive satisfaction from the consumption of final commodities or characteristics. These final commodities are "nonmarket basic commodities." Leisure time, for example, is an element of the bundle of final commodities in Becker's (1965) model. Nonleisure time used in the production of the basic commodities is assumed, implicitly, to have no contribution to the utility of the household. The production technology is usually assumed nonjoint, and the production of the basic commodities depends on the market goods, labor, and a given stock of capital.

MARKET DEMAND SYSTEMS

The demand systems reviewed have been applied in market contexts. Market or aggregate contexts are differentiated from individual contexts on the basis of whether the data used for estimation are from individuals or from aggregates of individuals. Generally, the reconciliation of individual consumer behavior with estimated functions based on market data requires behavioral restrictions of the types reviewed in connection with the specific demand systems.

The general theory of market demand functions is well established (Gorman 1953, 1956, 1976; Muellbauer 1975; Shafer and Sonnenschein 1982). These results are primarily impossibility theorems (Sonnenschein 1973a, 1973b). Specifically it has been shown by Sonnenschein that proceeding from the individual demand specifications, equation (4.2), there are no restrictions on the market demand except homogenity and budget equality. These results have been generalized somewhat by Debreu (1974), Diewert (1977), and Mantel (1977). But, the general conclusion from these assessments of the implications of the individual theory for market specifications is that the demand systems in market context exhibit few restrictions from the individual theory unless the utility functions of consumers are highly restricted (Geanakoplos and Polemarchakis 1980; Eisenberg 1961; Chipman and Moore 1976). The more recent results modestly generalized earlier conclusions, and show that the income distribution does not have to be fixed and that the Engel curves do not have to be linear and parallel for aggregation to occur. But these results are at present of limited use for linking individual and market demands (Hildenbrand 1983; Chiappori 1985).

An alternative approach to the specification of market demand systems has been taken by Johnson et al. (Johnson et al. 1986; Huang 1985; Haidacher 1983; Safyurtlu, Johnson, and Hansen 1986; and Byron 1984). This approach involves the idea of approximation or local approximations to market demand systems. The traditional results on market demand are primarily for global approximations. Approximation has been directly investigated by Byron (1984) for the Rotterdam model, showing conditions under which the bias generated by the approximation can be significant. Mountain (1988) showed that the discrete Rotterdam formulation can be derived from an approximation in quantities, prices, and income. He argued that since discrete formulation is an approximation of the corresponding variables, the estimated elasticities are also approximations. The implication is that for available data, approximations may be justified on the basis of the types of measurements that are used in generating the variables used for estimating the demand systems.

More recent extensions have involved further refinements of the ideas most prominently illustrated in the work of Byron (1984). That is, demand systems have been specified as approximate and rationalized for use in individual and market contexts. Practical results from this work are limited. A new alternative for investigating the nature of the approximations and their local accuracy uses Monte Carlo or numerical methods. Results to date indicate that relatively simple forms of demand functions performed as well as more complex specifications as approximations of underlying consumer behavior if applied in synthesized market data. That is, it appears that there is little gain from applying complex demand specifications in market data, if the argument is that these specifications are only approximations of the aggregate consumer behavior (Kesavan 1988; Johnson and Kesavan 1988).

CONCLUSION

Increasingly, demands for agricultural and food commodities are being estimated and evaluated in a systems context. This is because of the recognized importance of substitution in the demand specifications and empirical experience that suggests that these substitution effects are highly dependent on the demand specification used in the applied work. Using demand systems with more solid microfoundations provides a rationale for choice and testing that has not been available for the ad hoc functions typical of applied work in food and agricultural demand analysis. But, the added rigor and "improved" specifications come at a cost. The old adage holds, to get something you have to assume something. The question then is whether or not the assumptions required to apply these systems can be rationalized on the basis of the available data and observed consumer behavior.

Major problems for applications of modern demand theory for food and agricultural commodities involve the fact that the available data are from markets. Much of the theory of individual consumer demand does not carry over to the market level. That is, the restrictions that underlie the demand systems and provide testable hypotheses at the individual level do not apply in market data without highly limiting assumptions on the utility function.

Two approaches have been prominent in addressing this problem. The first has investigated demand systems that proceed from restricted utility functions and that apply in market data; the linear expenditure system, the AIDS, and the Rotterdam model are examples. Experience in applying these systems to date suggests they are more appropriate for aggregated commodities. This is not encouraging for agricultural and food demand analysis since most policy and forecasting problems require highly disaggregated demand functions.

The second approach is more pragmatic. That is, simple systems have been applied in estimating market demand parameters and for disaggregated commodity specifications. The argument has been that these demand systems somehow approximate the underlying consumer behavior. But, the characteristics of the "approximation" have not been established. Recently, this approach has received increased attention from analysts. Experiments have been conducted using computer simulation techniques to identify the nature of the approximations implied by the simple demand functions, and the adequacy with which consumer behavior can be approximated by such functions estimated in market data. The results of this work have been encouraging, suggesting improved foundations for demand estimates from market data.

The theory and empirical work on the demand systems provides a rich basis for investigating consumer behavior and developing market level parameters that can be applied to forecasting policy analysis for the dairy industry. Analysts are rapidly moving to take advantage of these results from the theory in their empirical work. More complete demand specifications are being developed, and the basis for interpreting the empirical results is being improved, leading to added understanding of the demand for dairy products and an improved basis for a development of policy for the dairy industry.

REFERENCES

Bartens, A. P 1969. "Maximum Likelihood Estimation of a Complete System of Demand Equations." *European Economic Review* 1:7–73.

_____. 1977. "The Systems of Consumer Demand Functions Approach: A Review." *Econometrica* 45:23–51.

Becker, Gary S. 1965. A Theory of the Allocation of Time. *Economic Journal* 75:493–517.

Berndt, E. R. and Mohammed S. Khaled. 1979. "Parametric Productivity Measurement and Choice Among Flexible Functional Forms." *Journal of Political Economy* 87:1220–1245.

Box, G. E. P. and D. R. Cox. 1964. "An Analysis of Transformations." *Journal of the Royal Statistical Society* 26:211–243.

Brown, M., Richard Green, and S. R. Johnson. 1986. Preferences and Optimization Tests Within the Almost Ideal Demand System.

Byron, R. P. 1984. "On the Flexibility of the Rotterdam Model." *European Economic Review* 24:273–283.

Chiappori, Pierre-Andre. 1985. "Distribution of Income and the Law of Demand." *Econometrica* 53:109–127.

Chipman, J. S. and J. Moore. 1976. On the Representation and Aggregation of Homothetic Preferences. Manuscript.

Christensen, L. R., D. W. Jorgenson, and L. J. Lau. 1975. "Transcendental Logarithmic Utility Function." *American Economic Review* 65:367–383.

Deaton, A. and J. Muellbauer. 1980a. *Economic and Consumer Behavior*. Cambridge: Cambridge University Press.

_____. 1980b. "An Almost Ideal Demand System." *American Economic Review* 70: 312–326.

Debreu, G. 1974. "Excess Demand Functions." *Journal of Mathematical Economics* 1:15–23.

Diewert, W. E. 1977. "Generalized Slutsky Conditions for Aggregate Consumer Demand Functions." *Journal of Economic Theory* 15:353–362.

Eisenberg, B. 1961. "Aggregation of Utility Functions." *Management Science* 7:337–350.

Engel, E. 1895. Die Lebenskosten Beligisher Arbeiter-Families früher und jetzt. International Statistical Institute Bulletin 9:1–74.

Frisch, R. 1959. A Complete Scheme for Computing All Direct and Cross Demand Elasticities in a Model with Many Sectors. *Econometrica* 27:177–196.

Geanakoplos, J. D. and H. M. Polemarchakis. 1980. "On the Disaggregation of Excess Demand Functions." *Econometrica* 48:315–331.

Goldberger, A. S. 1969. "Directly Additive Utility and Constant Marginal Budget Share." *The*

Review of Economic Studies 36:251–254.

Gorman, W. M. 1953. "Tricks Community Preferences Fields." *Econometrica* 21:63–80.

_____. 1956. "A Possible Procedure for Analyzing Quality Differentials in the Egg Market." *Review of Economic Studies* 47:843–856.

_____. 1976. "Tricks with Utility Functions." In *Essays in Economic Analysis*, eds. M. Artis and R. Nobay. Cambridge: Cambridge University Press.

Haidacher, R. D., J. A. Craven, K. S. Huang, D. M. Smallwood, and J. R. Blaylock. 1982. *Consumer Demand for Red Meats, Poultry, and Fish*. Washington DC: U.S. Department of Agriculture, Economic Research Service.

Hendler, R. 1975. "Lancaster's New Approach to Consumer Demand and Its Limitation." *American Economic Review* 65:194–199.

Hildenbrand, W. 1983. "On the 'Law of Demand.' " *Econometrica* 51:997–1020.

Houthakker, H. S. 1960. "Additive Preferences." *Econometrica* 28:244–257.

Huang, K. S. 1985. "U.S. Demand for Food: A Complete System of Price and Income Effects." Technical Bulletin No. 1714, United States Department of Agriculture, Economic Research Service.

Johnson, S. R., Z. A. Hassan, and R. D. Green. 1984. *Demand Systems Estimation—Methods and Applications*. Ames: Iowa State University Press.

Johnson, S. R., R. D. Green, Z. A. Hassan, and A. N. Safyurtlu. 1985. Market Demand Functions. Mimeo.

Johnson, S. R., R. Green, Z. Hassan, and A. Safyurtlu. 1986. "Market Demand Functions." In *Food Demand Analysis: Implications for Future Consumption*, eds. Oral Capps Jr. and Benjamin Senauer. Virginia Polytechnic Institute and State University, Blacksburg.

Johnson, S. R. and T. Kesavan. 1988. "Progress Report: Monte Carlo Experiments of Market Demand Theory." Progress Report for Cooperative Agreement #58-3AEK-7-00044 between Center for Agricultural and Rural Development and the Economic Research Service, Commodity Economic Division, USDA (September).

Kesavan, Thulasiram. 1988. Monte Carlo Experiments of Market Demand Theory. Iowa State University, Ames.

Lancaster, Kelvin J. 1966. "A New Approach to Consumer Theory." *Journal of Political Economy* 74:132–157.

Lucas, Robert E. 1975. "Hedonic Price Function." *Economics Inquiry* 13:157–178.

Mantel, R. 1977. "Implications of Microeconomic Theory for Commodity Excess Demand Functions." In Vol. IIIA of *Frontiers of Quantitative Economics*, eds. M. D. Intriligator, Amsterdam: North Holland.

Mears, L. A. 1981. The New Rice Economy of Indonesia. Jakarta: Gadjah Mada University Press.

Michael, R. T. and G. S. Becker. 1973. "On the New Theory of Consumer Behavior." *Swedish Journal of Economics* 75:378–396.

Muellbauer, J. 1975. "Aggregation, Income Distribution and Consumer Demand." *Review of Economic Studies* 62:525–543.

Muth, Richard F. 1966. "Household Production and Consumer Demand Functions." *Econometrica* 34:698–708.

Phlips, L. 1983. *Applied Consumption Analysis*. Amsterdam: North Holland.

"Policy Issues, Project Design and Implementation Plan: Final Research Strategies Based on Host Country Review." Report #1-Indonesia. Center for Agricultural and Rural Development, Iowa State University. July 1985.

Pollak, R. A. 1971. "Additive Utility Functions and Linear Engel Curves." *Review of Economic Studies* 38:401–413.

_____. 1978. "Estimation of Complete Demand System from Household Budget Data: The Linear and Quadratic Expenditure Systems." *American Economic Review* 68:348–359.

_____. 1979. "Welfare Comparisons and Equivalence Scales." *American Economic Review*

69:216–221.
____. 1981. "Demographic Variables in Demand Analysis." *Econometrica* 49:1533–1555.
Prais, S. J. and H. S. Houthakker. 1955. *The Analysis of Family Budgets*. Cambridge: Cambridge University Press.
Safyurtlu, A. N., S. R. Johnson, and Z. A. Hassan. 1986. "Recent Evidence on Market Demand Systems for Food in Canada." *Canadian Journal of Agricultural Economics* 34:475–493.
Shafer, W. and H. Sonnenschein. 1982. "Market Demand and Excess Demand Functions." In Vol. II. of *Handbook of Mathematical Economics*, eds. K. J. Arrow and M. D. Intriligator. Amsterdam: North Holland.
Sonnenschein, H. 1973a. "Do Walras' Identity and Continuity Characterize the Class of Community Excess Demand Functions?" *Journal of Economic Theory* 6:345–354.
____. 1973b. "The Utility Hypothesis and Market Demand Theory." *Western Economic Journal* 11:404–410.
Stone, J. R. N. 1954. Linear Expenditure Systems and Demand Analysis: An Application to the Pattern of British Demand. *Economic Journal* 64:511–527.
Teklu, Tesfaye. 1984. An Economic Analysis of Resource and Income Uses Among Farm Households of Ethiopia: Applications of the Household Production Approach. Ph.D. dissertation. Iowa State University, Ames.
Theil, H. 1965. "The Information Approach to Demand Analysis." *Econometrica* 33:64–87.
____. 1969. "A Multinomial Extension of the Linear Logit Model." *International Economic Review* 10:251–259.
____. 1971. *Principles of Econometrics*. New York: John Wiley & Sons.
Timmer, P. C. 1971. "Estimating Rice Consumpion." *Bulletin of Indonesian Economic Studies* (July).
Timmer, P. C. and H. Alderman. 1979. "Estimating Consumption Parameters for Food Policy Analysis." *American Journal of Agricultural Economics* 61.
Tyrrell, T. and T. Mount. 1982. "A Nonlinear Expenditure System Using a Linear Logit Specification." *American Journal of Agricultural Economics* 64:539–546.
Varian, H. R. 1978. *Micro Economic Analysis*. New York: W. W. Norton.
World Bank. 1983. "Indonesia: Policy Options and Strategies for Major Food Crops." Report No. 36865-IND (April 4)
Yoshihara, K. 1969. "Demand Functions: An Application to the Japanese Expenditure Pattern." *Econometrica* 37:257–274.

SUGGESTED FOR FURTHER READING

Barten, A. P. 1964. "Consumer Demand Functions Under Conditions of Almost Additive Preferences." *Econometrica* 32:1–38.
Boedino. 1978. "Elasticitas Permintaan Untuk Ber bagai Barang di Indonesia: Penerapan Metode Frisch." *Ekonomi dan Keldangan Indonesia* (September).
Brown, J. A. C. and A. Deaton. 1972. "Surveys in Applied Economics: Models of Consumer Behavior." *The Economic Journal* 82 (December):1145–1236.
Chernichovsky, D. and D. A. Meesook. 1984. Patterns of Food Consumption and Nutrition in Indonesia. World Bank Staff Working Paper #670 (September).
Dixon, J. A. 1982. "Food Consumption Patterns and Related Demand Parameters in Indonesia: A Review of Available Evidence." Mimeo.
Hedley, Douglas D. 1978. Supply and Demand for Food in Indonesia. Paper presented at the Workshop on Research Methodology (March), Department of Agricultural Economics and Rural Sociology, Bogor Agricultural University.
Ladd, George W. and Veraphol Suvannut. 1976. "A Model of Consumer Goods Characteris-

tics." *American Journal Agricultural Economics* 58:504–510.

LaFrance, Jeffrey T. 1983. The Economics of Nutrient Content and Consumer Demand for Food. Ph.D. dissertation. University of California, Berkeley.

Lau, L. J. and B. M. Mitchell. 1975. "A Linear Logarithmic Expenditure System: An Application to U.S. Data." *Econometrica* 65:367–383.

Magira, Stephen L. 1981. "The Role of Wheat in Indonesia's Food System." Foreign Agricultural Economic Report No. 170 (December), USDA.

Muellbauer, J. 1974. "Household Production Theory, Quality, and the 'Hedonic Technique.' " *American Economic Review* 64:977–994.

Parks, R. W. 1969. "System of Demand Equations: An Empirical Comparison of Alternative Functional Forms." *Econometrica* 37:629–650.

Pollak, R. A. and M. C. Watcher. 1975. "The Relevance of the Household Production Function and Its Implications for the Allocation of Time." *Journal of Political Economy* 83:255–277.

Pollak, R. A. and T. J. Wales. 1969. "Estimation of the Linear Expenditure System." *Econometrica* 37:611–628.

Singh, I, L. Squire, and J. Strauss. 1985. *Agricultural Household Models: A Survey of Recent Findings and Their Policy Implications*. Economic Growth Center: Yale University.

Teklu, Tesfaye. 1984. An Economic Analysis of Resource and Income Uses Among Farm Households of Ethiopia: Applications of the Household Production Approach. Ph.D. dissertation, Iowa State University, Ames.

Theil, H. 1965. "The Information Approach to Demand Analysis." *Econometrica* 33:64–87.

Timmer, P. C. 1971. "Estimating Rice Consumption." *Bulletin of Indonesian Economic Studies* (July).

Tyrrell, T. and T. Mount. 1982. "A Nonlinear Expenditure System Using a Linear Logit Specification." *American Journal of Agricultural Economics* 64:539–546.

World Bank. 1983. "Indonesia: Policy Options and Strategies for Major Food Crops." Report No. 36865-IND (April 4).

CHAPTER 5

Quantitative Requirements of a Supply-Managed Industry: The Case of the Canadian Dairy Industry

Merritt E. Cluff and D. Peter Stonehouse

INTRODUCTION

The policy organization of the Canadian dairy industry offers a unique setting in which quantitative methods in general and policy modeling in particular may be useful. The interrelationship of the information requirements of supply management and the services provided by industry/policy models is useful to explore, as is a discussion of the strengths and weaknesses of these tools.

Supply management of the Canadian dairy industry is, from an agricultural perspective, a classic case of a regulated quasi-public monopoly. The split in jurisdiction of the fluid (fresh table) milk and industrial milk (for cheese, butter, and other manufactured products) sectors between the provincial and federal governments complicates a uniform treatment of the sector in this monopoly paradigm, but the operation of supply management in the industrial sector provides ample evidence whereby specific and very direct policy instruments are used to achieve a reasonably well-defined set of objectives given a set of constraints. Modeling in this context covers a wide domain in the quantitative economic policy literature. On the one hand, supply management involves the notion of forward planning concerning price and quantity levels and also producer and public

---

Senior authorship was not assigned. The views expressed in this chapter are those of the authors and not necessarily those of Agriculture Canada or of the University of Guelph. Assistance from Elisha Kizito, Agriculture Canada, is gratefully acknowledged.

expenditure levels. Early attempts at modeling the Canadian dairy industry emphasized these aspects (Stonehouse et al. 1978). On the other hand, supply management entails some notion of an objective function for the industry subject to perceived or legislative constraints; this raises modeling applications as suggested by Tinbergen (1955) and also by Fox et al. (1966), not to mention more recent literature on optimal control. For both these applications, the issue of the quality of model information arises, which in turn relates to the ability to identify and estimate important parameters or to quantify target variables as well as model policy behavior and constraints (Rausser and Stonehouse 1978; Stonehouse and Rausser 1981).

This chapter attempts to explore the application of an industry/policy model to the case of the supply-managed industrial milk sector in Canada. The next section briefly discusses the information needs of supply management. This is followed by a section outlining the information provided by industry/policy models; in this section the relationship among policy objectives, policy instruments, and model structure is explored, noting important parameters that need to be estimated. The subsequent section outlines basic empirical evidence from a particular model currently in use at Agriculture Canada. A final section concludes the chapter.

INFORMATION REQUIREMENTS OF SUPPLY MANAGEMENT

Supply management is an institutional tool designed to bring order and control to the supply side of competitive industries, such as agriculture, characterized by an atomistic structure, relatively inelastic demand with respect to prices and incomes, and rapid, pervasive technological change. Such industry characteristics can lead to destructive competition among producers and, in the face of an oligopsonistic processing sector (in particular with nonstorable farm commodities), can lead to market failure. Supply management is a method of supplanting the discipline of the open market price-setting mechanism with a system for regulating the aggregate amount and temporal flow of a product reaching the marketplace. By controlling the supply side, market sources of risk can be reduced, laying the groundwork for achievement of supply management objectives, namely the raising and stabilizing of producer gross incomes. The achievement of these objectives involves significant information requirements on the part of the institution regulating the market. Such regulation, or management, implies the need for planning/forecasting of a broad range of variables, most notably aggregate demand, or quota requirements. It also implies the need for responsible management of the instruments afforded to it in support of its goals. Both these needs require good descriptive information on the economic, policy, and technical structure of the industry.

FORECASTING NEEDS

In terms of operational planning or forecasting, information needs stem directly from the supply-management arrangements that replaced the open market system in the mid-1960s. The policies and institutional programs are designed to ensure that domestic demand for high-quality products is fully met on a year-round basis at prices that assure efficient producers a reasonable return to resource inputs (Government of Canada 1965). Setting producer prices at an appropriate level to provide desired rates of return is the base point for initiating the process of assigning values to the various policy instruments for the next planning period. (This planning period is typically a dairy year of August to July, but since dairy policy has been reviewed each five years, it also has a longer term dimension.)

The setting of the national "target return" (Stonehouse 1987, 52) is guided by the notion of "fair return for labor and investment" and has been determined by different methods since 1966. Currently, cost surveys are used to establish the return. However, the federal government's dairy policy agency, the Canadian Dairy Commission (CDC), is not empowered to set producer milk prices directly because that is a provincial government jurisdiction. The CDC's mandate is, given a pricing mechanism established and approved by the federal government, to purchase key processed dairy products (butter, skim milk powder, and cheddar cheese) surplus to market requirements at guaranteed minimum prices. These offers to purchase establish a price floor at the wholesale level for all milk products given their relationship to the supported commodities. Two important sets of information are then required; the first involves relating wholesale prices backward to producer prices for milk, and the second, relating wholesale prices forward to retail prices and demand.

Relating wholesale prices to producer prices requires knowledge of conversion rates of milk into the various processed products and the marketing margins associated with processing plants. With this knowledge, the CDC is able to determine the support prices necessary to achieve a given target return at the producer level. Then marketing boards in the individual provinces use the target return to set minimum producer prices within their jurisdictions for milk according to its intended end use. Consequently, in terms of planning information, it is the target price that most needs to be forecast, since provincial adjustments follow, and along with the target price the support prices required to ensure that the target is met.

Relating wholesale prices forward to retail prices, and the consequences of these retail prices for demand, requires knowledge of wholesale-retail marketing margins, the effects of retail price changes on quantities demanded (demand elasticities, own and cross) of each dairy product, and other factors (income, taste trends, advertising effects, and so forth) that influence demand. Estimates of demand in turn form the basis for setting national quota levels for industrial milk supplies. Such restrictions are necessary because Canadian wholesale level support prices and associated producer target returns and retail prices are typically higher than prices of internationally traded dairy products, so that only domestic demand requirements can justifiably be filled from domestic sources of milk. Restrictions on imports of dairy products and any potential exports must also be factored into these considerations.

In summary, the establishment of producer target returns for industrial milk forms the logical starting point in setting policy instruments for the next planning period. This enables wholesale and retail prices to be forecast, together with demands for dairy products. The forecast summation of demands for dairy products provides a basis for setting upper limits on domestic milk production levels needed to meet domestic demand after allowing for net international trade movements. Forecasting demand for the Canadian dairy industry is not done in isolation but rather as an integral part of forecasting prices, supplies, stocks, trade movements, and key policy variables. In addition, the estimates provide a basis for other estimates such as government costs, producer levies, and so on.

Policy Evaluation Information Needs

There exists an extensive literature on the economic effects of supply management relative to an open market framework (see, for example, Forbes et al. 1982; Josling 1981; Veeman 1982). However, less work has been accomplished on the evaluation of the operation of supply management, how well it has achieved its objectives, how it can minimize economic losses and such. Given the existence of supply management, the setting of dairy policy instruments has direct implications for the economic environment of the sector. The main policy levers, such as the setting of support prices, subsidies, levies, and quota, require similar knowledge of the economic relationships that are required for forecasting, but in addition, the delineation of objectives is also required, not only in terms, of producer net incomes and price stability but also in terms of domestic economic efficiency and international trade concerns. In an operational sense, revision through time of the basic instruments may be minor, but periodic reassessment of issues such as the basis for quota determination, price levels, and allocation of support prices is an important requirement of public policy.

INDUSTRY/ECONOMIC POLICY MODEL INFORMATION

The role of quantitative models in planning, forecasting, and policy evaluation has been clearly established in the economics literature as well as in government and industry operations. An important aspect of this role is the nature of information provided, its value, and the critical parameters that determine these. A recent working paper (Stonehouse and Kizito 1990) outlines a detailed specification of a model of the Canadian dairy sector, which ties policy instruments to important aggregate variables in the industry. This quarterly econometric model serves as a basis for the following discussion, and for pedantic purposes the model described below is cast in a policy-oriented perspective.

The basic economic functioning of the milk sector in Canada can be described by Figures 5.1 and 5.2, where the former outlines industry equilibrium production, consumption, and price of both fluid and industrial milk, and where the latter outlines the equilibrium for industrial milk with respect to the two basic milk components, namely butterfat and nonfat solids. The model described here by equations 5.1 to 5.13 permits a generalized analysis of the linkage of policy instruments to objectives in the context of current Canadian industrial milk policy.

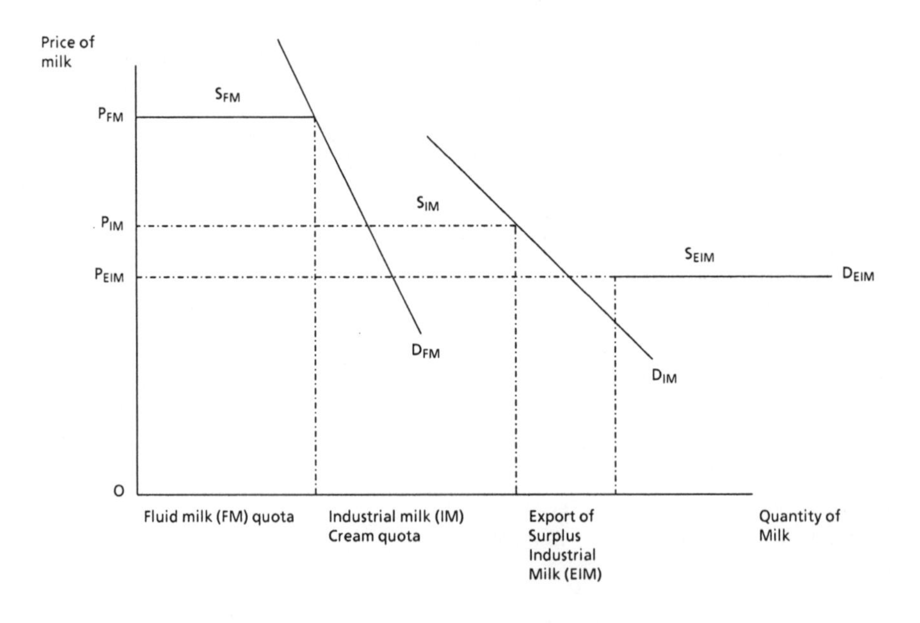

Figure 5.1. Equilibria in the supply-managed Canadian dairy industry

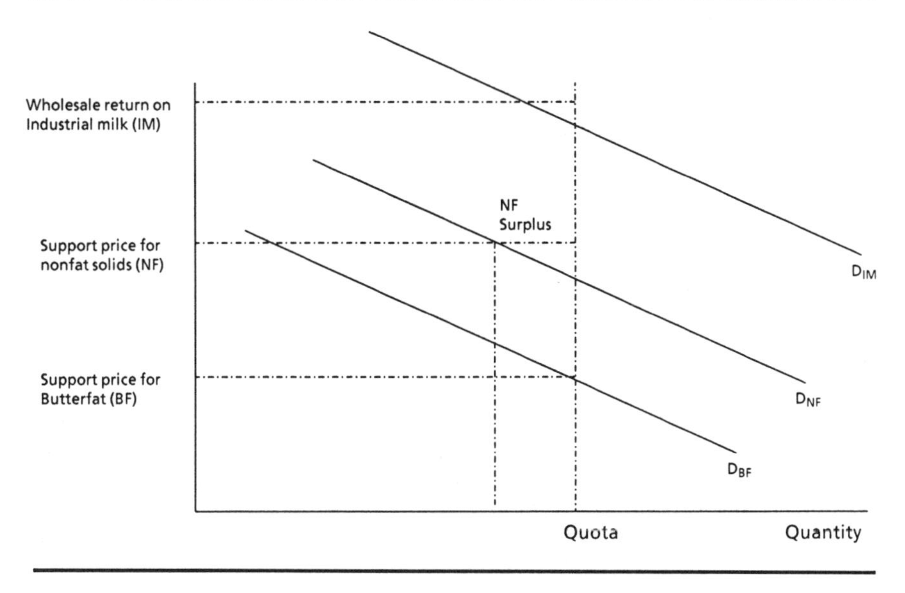

Figure 5.2. Supply and demand by industrial milk components, butterfat and nonfat solids

$P_i = COP$	(P_i - target return)	(5.1)
$Q_i = D_b/a_i$	(Q_i - quota)	(5.2)
$D_b = f(P_b, P_o, Y)$	(D_b - butterfat demand)	(5.3)
$D_{nf} = f(P_{nf}, P_p, Y)$	(D_{nf} - solids-not-fat demand)	(5.4)
$S_i = Q_i$	(S_i - industrial milk supply)	(5.5)
$S_b = a_1 S_i$	(S_b - butterfat supply)	(5.6)
$S_{nf} = b_1 S_i$	(S_{nf} - solids-not-fat supply)	(5.7)
$I_b = S_b + I_{b,t-1} + M_b - X_b - D_b$	(I_b - butterfat stocks)	(5.8)
$I_{nf} = S_{nf} + I_{nf,t-1} + M_{nf} - X_{nf} - D_{nf}$	(I_{nf} - solids-not-fat stocks)	(5.9)
$P_i = XS + P_b/a_1 + P_{nf}/b_1 - PM$	(basic price linkage)	(5.10)
$L_b = X_b(P_b - WP_b)$	(L_b - butterfat disposal costs)	(5.11)
$L_{nf} = X_{nf}(P_{nf} - WP_{nf})$	(L_{nf} - s_{nf} disposal costs)	(5.12)
$NR_i = P_i S_i - L_b - L_{nf} - C(S_i)$	(NR_i - net revenue, profits)	(5.13)

where

COP	is cost of production,
P_b	is support price butterfat (butter),
P_{nf}	is support price solids-not-fat (skim milk powder),
P_o	is price of other oils,
P_p	is price of other proteins,
Y	is income,

a_1	is butterfat content of milk,
b_1	is solids-not-fat content of milk,
M_b	is imports of butterfat,
M_{nf}	is imports of solids-not-fat,
X_b	is exports of butterfat,
X_{nf}	is exports of solids-not-fat,
WP_b	is world price butterfat,
WP_{nf}	is world price solids-not-fat,
XS	is government subsidy to producers,
PM	is assumed processors' margin, and
$C(S_i)$	is cost function, industrial milk.

In this model the important policy instruments are potentially numerous, but in actual operation, conventions have largely restricted choices. Equation (5.1) indicates that target returns, a proxy for producer prices, are regulated by an exogenous formula related to costs of production (average for a specific group of producers in $/hectoliter). This relationship constitutes the major constraint to this monopoly-oriented industry. Equation (5.2) specifies quota levels at the domestic demand level for butterfat, while equations (5.3) and (5.4) specify conventional demand functions for the basic milk products. Equations (5.5) to (5.7) specify milk supply as equal to the set quota level, and the supplies of butterfat and nonfat solids corresponding to this level of supply.

Equations (5.8) and (5.9) reflect the offer-to-purchase instrument by which, at prices P_b and P_{nf}, commodities are purchased by a central agency (the Canadian Dairy Commission); in this case, public stocks move according to demand and supply conditions with offsetting changes made by the central agency to import or export a product if a desired level of inventory is to be attained. By convention, with quota set at demand levels for butterfat, levels of exports and imports of butterfat are expected to be in balance and can be assumed to be exogenous, and stock changes can be assumed to be nil. In practice, sustained changes in butterfat stock levels are leading indicators of required quota changes. The same restrictions do not apply for exports and imports of solids-not-fat, and depending on the prices, these levels may differ substantially in practice with exports vastly exceeding imports.

Equation (5.10) is a fundamental price equation, linking product prices at wholesale, processors margins (PM), a government subsidy (XS), and producer prices. It is important to note that with PM and XS exogenous and with P_i given from equation (5.1), there is no formal mechanism for determining the relative butterfat (B_f) and solid nonfat (S_{nf}) support prices, although latent in the choice of these prices through time may be a number of factors related to image, public acceptance, and such. During the 1970s increases in support prices required to achieve changes in indicated target levels of P_i were made by increasing P_{nf} relative to P_b since such a strategy preserved market size measured in terms of butterfat (see Fig. 5.3). In the 1980s, a greater tendency has been to assign increases on a fifty-fifty basis to the support prices. Equation (5.13) provides a calculation of net industry income equal to total revenue less surplus disposal cost less costs of production.

In practice, the policy instruments in this model have been severely constrained. For example, the determination of COP to provide "a fair return for labor and investment" has been a contentious issue since its proposal in 1975; this level is a policy choice, at least until the process generating the COP has been chosen. By similar token, choosing the definition of quota (Q_i) has been a policy choice decision. Similarly, constraints are placed on imports and exports of butterfat and imports of nonfat solids. Operatively, exports of nonfat solids are set to reduce stocks to desired levels, and hence while the central agency controls exports, stock levels are effectively exogenous and exports endogenous to the system. Constraints on support price determination as noted above also apply. Policy choice, even with respect to changes in subsidy levels (XS) and processor margin levels (PM) has been limited by political/interest group factors.

The normal operation of industrial milk supply management has been to update prices and develop forecasts of demand levels (that is, for quota in terms of butterfat) and appropriate levies to producers. Hence the information requirements for supply management as supplied by a model such as that described is to produce the conditional forecast for these variables, given expected movements in the exogenous factors driving the system (5.1) to (5.13), that is, cost of production (COP), price of other oils (P_o), price of other proteins (P_p), income (Y), trend variable (T), world price of butterfat (W_{pb}), and world price of nonfat solids (W_{pnf}). However, forecasting required quota levels in the short term has been limited; temporary changes in quota (Q_i) may significantly affect holders of quota in that, with constrained output, quotas have value that could be destabilized by uncertainty as to quota volume changes. The central agency has preferred to vary butter stock levels in the short term to stabilize quota levels, or to offer "butter sales" to reduce inventory. In this sense, forecasting certain key industry variables such as quota levels may be more useful as an indicator of general directions and less useful in current decisions with respect to these policy parameters. However, the model potentially provides conditional forecasts with detail useful for analyzing the interrelationships forecasts within the industry over a forecast period based on the set of estimated parameters.

The main strength of this model is its use in a programming, or control, application in the analysis of supply management or in the testing of alternative regimes. For instance, consider the problem of maximizing producer incomes with respect to alternative policy parameters, subject to specific constraints. For example, consider system (5.1) to (5.13), with (5.13) as an objective function (i.e., maximize net incomes of producers) and/or with alternative constraints imposed such as minimizing variations in quota levels or choosing different price allocations to P_b and P_{nf} given a fixed price P_i. In this case, the estimated model can be used as a programming model in which the optimum choices of the policy instruments can be searched for so that the objectives can be achieved most efficiently. In a theoretical sense, with a knowledge of the parameters of the system described, one could solve for the marginal conditions at the optimum sought and provide analytical information in terms of the levels at which the policy instruments might be set to achieve these. A practical example of this problem was the use of the butterfat exchange program of the late 1970s. In this case butter was imported and certain other products were exported to improve net incomes of producers. This was made possible by the fact that domestic butter and skim milk powder prices were out of line compared to international prices for certain products, which at the time mainly included evaporated milk.

In using model information for evaluative purposes in terms of interest to domestic producers, information on the cost function $C(S_i)$ or supply function is required. However, in the presence of supply management, it has not been possible to estimate this function econometrically, and the marginal cost schedule is not obtainable. As a result, without this knowledge, calculation of the level of instruments for an optimum net revenue (NR_i) is not possible explicitly. It nevertheless remains a useful analytic concept, and indicators such as returns over cash costs may offer a substitute in examining program options. An example of this is provided below.

EMPIRICAL EXPERIENCE WITH AN ECONOMETRIC MODEL

In satisfying the quantitative needs of supply management, the issues of accuracy/reliability and applicability naturally arise. Measuring the value of information from a model is a difficult task. Forecasting performance can be assessed from experience (ex post or ex ante errors) and these can be compared to naive forecasting methods or to numerous other approaches. More difficult to assess is a model's ability to perform evaluative exercises in counter-factual or hypothetical circumstances such as assessing the impact of alternative program designs. In such a case, statistical information on key parameters and actual experience with the model provides evidence of reliability. In terms of applicability, the question of whether the model can address specific real issues can be explored by comparative simulations and by assessing the realism of the results. In this section, the parameters critical to the dairy model described above and the issues that affect their estimation are first examined. Selected indicators of forecast performance are then presented along with a discussion of alternative forecasting techniques. Finally, the evaluation of specific policy alternatives is provided.

Model Parameters and Estimation Issues

The critical parameters of the system described relate to the demand equations D_b (5.3) and D_{nf} (5.4), as well as the net income equation NR_i (5.13), specifically the estimation of cost function $C(S_i)$. The quality of forecast information on demand requirements is directly dependent on the accuracy of the demand equation estimates. As well, these estimates necessarily affect the ability of the model to assess the impact of alternative policy instruments on the welfare of consumers and producers.

Table 5.1 provides recent estimates of the elasticities for major milk products. A characteristic of these estimates is the difficulty in estimating the elasticities of demand for butter and for skim milk powder. In fact, through the 1970s and the 1980s, the robustness of these estimates has deteriorated substantially to the point where time-series estimates of the demand function for butter have broken down entirely. The reasons appear to lie primarily with a dominating secular trend toward decreased dietal preferences for butter and saturated animal fats generally. However, a further problem relates to experimental design problems in the time-series estimation of this function. This problem arises from the increased stability of real dairy prices in general and butter price in particular since the mid-1970s when milk prices were increased substantially with the inception of effective supply management. The issue is that actual variation of real butter prices has been minimal in comparison with total variation in per capita quantities; the result is that the observation period 1975–1988 offers little evidence from which to choose the own-price elasticity. Estimations reveal substantial multicollinearity in the variables explaining butter consumption, in particular between the butter price and the constant term of the relationship. As a result of this issue, constrained estimation of the butter demand function has been necessary, with constraints chosen largely from early work and from cross-sectional studies of food consumption.

Problems in estimating the dairy cost (and supply) function have been noted by Graham et al. (1990) among others. The issue in this context is that where quota is constraining, calculations of the industry cost curve separate from the value of quota are difficult. In addition, constrained output severs the relationship of output to price, thus impeding the identi-

Table 5.1. Elasticities of demand for dairy products

	Own-price	Cross-price	Income
Standard milk	−0.14	—	0.23
Low-fat milk	−0.32	—	0.25
Butter	−0.77	0.30 (margarine)	0.10
Skim milk powder	−0.08	0.91 (fluid milk)	1.58
Cheese	−0.57	0.07	0.88

Source: Recent estimates from dairy component of Food and Agriculture Regional Model. Butter cross-price and income elasticities have been constrained. Skim milk powder elasticities are insignificant.

ification of the industry supply curve. Without this knowledge, it is impossible to accurately gauge the impact on producers of alternative policies. In addition, forecasts of supply may also be questioned if prices fall below the marginal costs of production.

Forecasting Experience

The most reliable test of the performance of a forecasting system is the analysis of ex ante forecast errors, which include not only errors made given accurate projection of the surrounding environment of the industry but also the impact of unforeseen changes in this environment. Assessing a model in this setting is difficult since models may change over time and so may economic parameters. Records of forecasts may include results from different specifications, from different estimation periods, and from different personnel in their generation. Historical simulation is less useful since it assesses only inter-sample tracking performance; nevertheless it does provide information on model reliability given actual exogenous values, and is trivial to obtain.

Table 5.2 provides simulation errors from the dairy component of Agriculture Canada's FARM econometric model for the period 1986 quarter 3 to 1988 quarter 4. While these are not true forecast errors, they are indicative of those experienced over time. Prices are forecast reasonably well, although the extent of variation in prices is not high. Demand forecasts for both butter and cheese show reasonably low mean percentage errors, but high root mean squared percentage errors (variation). This latter result would seem to imply that the model may provide reasonable forecasts over a period of time but is relatively poor at forecasting particular quarters. In short, the ability of the model to forecast short-term activity would appear limited. This problem may be somewhat endemic since variation in apparent consumption of milk products is high. For more

Table 5.2. Mean percentage simulation errors, root mean squared errors for important variables, 1986 quarter 3 to 1988 quarter 4

	Mean percentage error	Root mean squared percentage error
Gross target returns	−1.8	2.3
Butter support price	−1.7	2.5
SMP support price	−1.5	2.2
In-quota levy	−26.6	27.7
Retail price index − dairy	−2.4	2.5
Disappearance − butter	1.5	7.4
− cheese	0.7	4.7
Market share quota	2.4	2.8
Market cash receipts	2.8	3.7

accurate forecasts, use must be made of other forecasting methods, such as expert opinion systems or time-series models (see Bates and Granger 1969, Leuthold et al. 1970, and Brandt and Bessler 1981 for comparisons of methods).

Accuracy of forecast results is, however, only one relevant criterion by which forecast users and decision makers can judge the choice of alternative. Other criteria are simplicity, versatility, and ease of updating. Moreover, alternative methods can be combined to provide so-called composite forecasts, which may incorporate information from the different methods. Maintaining a model that is rich in structure and supplementing its forecasting ability may provide an ideal instrument for decision makers wishing to forecast and analyze implications of alternative options at the same time.

Evaluative Simulations

Assessments of policy instruments constitute a potential use of models, but results need to be scrutinized fully to ensure not only that reasonable parameter values have been used but also that specifications are appropriate for the assessments in question. Presented below are the results of three policy simulations, which have been chosen as illustrative of the types of information that may be useful in policy assessment. Certain precautions are also noted.

Table 5.3 shows the effect of a 10 percent increase in the gross target returns under the normal operative assumption that such an increase is allocated on a fifty-fifty basis to the butter and skim milk powder support prices (neither the subsidy nor the processor margin was altered in this simulation). The assumption of a fifty-fifty basis increase to butter and skim milk powder support prices is consistent with the maintenance since 1974 of a constant ratio of butter and skim milk powder support prices (Fig. 5.3). The results indicate that such an increase in producer returns requires a similar rise in support prices, which in turn lowers market size in terms of butterfat by 6.6 percent on average over a two-year period. Net cash (gross receipts less cash costs) receipts rise by 36.5 percent. A major interesting feature of this simulation is that while higher skim milk powder prices lead to higher per unit disposal costs on the world market, higher butter prices cause lower demand and lower production of skim milk powder for export. The end result is a decrease in the in-quota levy to a level lower than what one might have expected.

These results, in conjunction with the earlier model discussion, raise the issue of the optimum allocation of support price levels to maximize producer incomes. Without knowledge of the cost function the solution of this problem is not possible; however, one can nevertheless explore the effect on net revenues of cash costs of support level adjustments. Table 5.4, for instance, shows the effect of a 20 percent increase in the skim milk powder support price on the important variables with the gross target price held constant. In this case, producers are made slightly better off in terms of gross cash receipts but not in net terms. Higher demand for milk in butterfat terms leads to higher sales, but higher butter production leads to higher powder production to be disposed of at higher per unit disposal costs. If costs of production increased in proportion to output, industrial milk producers would be less well off under such a change. Given this result, the reverse movement in support prices would increase producer net revenues, although market size would be substantially smaller.

Table 5.3. Selected multiplier effects for a 10 percent increase in the gross target returns for industrial milk

Selected variable	Multiplier %
Support price of butter	+10.4
Support price of skim milk powder	+ 9.1
Wholesale price of cheese	+ 8.2
Disappearance of butter	− 6.4
Disappearance of cheese	− 3.1
Industrial milk requirement	− 6.6
Exports of skim milk powder	−25.5
Industrial milk cash receipts (net of cash costs)	+36.5

Note: Results refer to the effect of the last four quarters of a ten-quarter simulation.

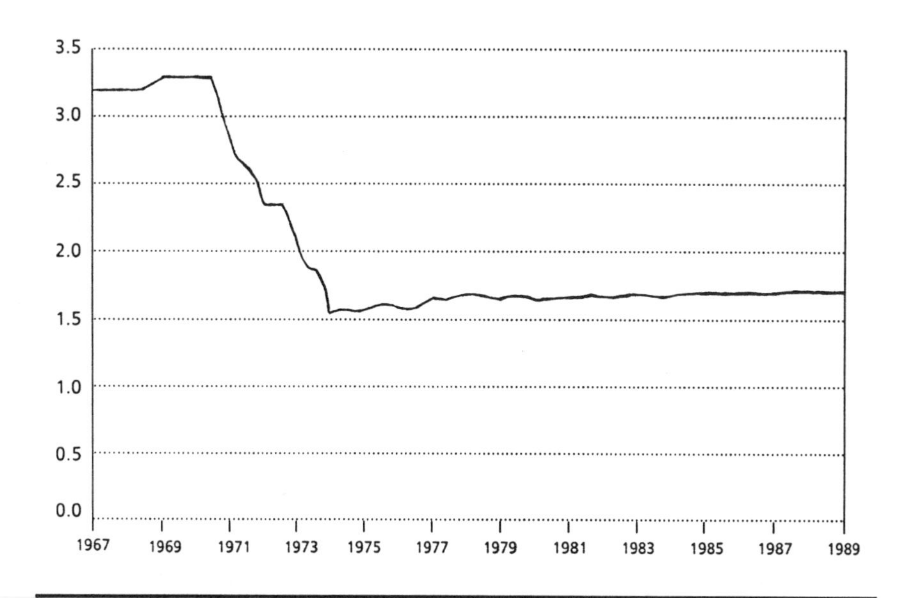

Figure 5.3. Butter to skim milk powder support price ratio

Table 5.4. Selected multiplier effects for a 20 percent increase in the skim milk powder support price

Selected variable	Multiplier %
Support price of butter	−22.6
Wholesale price of cheese	0.0
Disappearance of butter	+18.8
Disappearance of cheese	0.0
Industrial milk requirements	+13.6
Net target return	−6.5
In-quota levy	+107.0
Skim milk powder exports	+178.0
Industrial milk cash receipts (net of cash costs)	−13.6

Note: Results refer to the average effects of last four quarters of a ten-quarter simulation.

This experiment begs the question of where the model would find the optimum industry net receipts. This question was explored by iteratively simulating the model. Table 5.5 outlines results of a simulation in which skim milk product support prices have been reduced by 20 percent, with the butter support price increased to sustain the same gross target return.

Table 5.5. Optimizing industrial milk net cash receipts: a 20 percent decrease in the skim milk powder support price

Selected variable	Multiplier %
Support price of butter	+22.6
Wholesale price of cheese	0.0
Disappearance of butter	−12.8
Disappearance of cheese	0.0
Industrial milk requirements	−5.8
Net target return	+3.8
In-quota levy	−53.3
Skim milk powder exports	−63.1
Industrial milk cash receipts (net of cash costs)	+4.0

Note: Results refer to the average effects of last four quarters of a five-year simulation, 1984–1988.

A decrease greater or less than 20 percent resulted in lower industrial milk cash receipts although this measure is surprisingly "flat" over a wide range of changes to this support price. In this scenario, net cash receipts increased by 4 percent compared to the baseline, but a decrease in quota of almost 6 percent was involved. The results suggest that producers would be better off (although not significantly) with a higher butter price and a lower skim milk powder price because losses on skim milk powder exports would be substantially reduced. With significantly higher butter prices, however, consumers would be much worse off.

These types of evaluative simulations are useful for managers of supply-managed commodities as they provide useful information on the efficient setting of various policy parameters. Viewed in other terms, analysis of the functioning of supply management in terms of achieving stated goals is facilitated by quantitative tools that measure the effects of instruments on economic outcomes.

CONCLUSION

Decision makers participating in or responsible for regulated industries require three types of information: descriptive (for explaining the underlying structure of economic relationships), predictive (for guiding decisions about future events), and prescriptive (for evaluating the impacts of alternative policy choices). Econometric models are able to provide these types of information and do so on a consistent and systematic basis, combining the strengths of all three information sets. Information from the models may be supplemented with information from other sources.

In particular, econometric models offer descriptive information in the form of coefficient estimates for the parameters together with qualifying statistical tests, multipliers describing the impacts of changes to the exogenous variables in a model, and elasticity coefficients measuring the impact on an endogenous variable of a change in one of the model's parameters at a time. Predictive information can be generated by econometric models in the form of short-term or longer term forecasts or in the form of deterministic or stochastic forecasts. Prescriptive information may be obtained in the form of policy scenario comparisons or optimum policy paths according to predefined selection criteria.

The three types of information may be of assistance to decision makers operating in any of the market systems, but the needs may become critical for those responsible for policy setting in supply-managed or centrally planned systems. Econometric models tend, however, to be more complex to build, maintain and update, operate, and interpret than the more simplistic time-series models or the nonsystematic expert opinion systems. Econometric models also have more extensive and complicated data needs. However, from these models, decision makers can gain a better perspective on how to achieve their policy goals. In the case of the supply-managed Canadian dairy industry, the analysis of central agency operations assists managers in choosing instrument mixes that better serve the sector.

REFERENCES

Bates, J. M. and C. W. J. Granger. 1969. "The Combination of Forecasts." *Operation Research Quarterly* 20:451–468.

Brandt, J. A. and D. A. Bessler. 1981. "Composite Forecasting: An Application with U.S. Hog Prices." *American Journal of Agricultural Economics* 63(1):135–140.

Forbes, J. D., R. D. Hughes, and T. K. Warley. 1982. *Economic Intervention and Regulation in Canadian Agriculture*. Ottawa: Economic Council of Canada.

Fox, K. A., J. K. Sengupta, and E. Thorbecke. 1966. *The Theory of Quantitative Economics*. Amsterdam: North Holland.

Graham, J., B. Stennes, K. Meilke, G. Moschini, and R. MacGregor. 1990. *The Effects of Trade Liberalization on the Canadian Dairy and Poultry Sectors*. Working paper. Ottawa: Agriculture Canada.

Government of Canada. 1965. "An Act Providing for the Establishment of a Dairy Commission in Canada, 1966–67." Bill C34SL. Ottawa, Ontario: The Queen's Printer for Canada.

Josling, T. 1981. Intervention and Regulation in Canadian Agriculture: A Comparison of Costs and Benefits among Sectors. Technical Report No. 14. Ottawa: Economic Council of Canada.

Leuthold, R. A., A. MacCormick, A. Schmitz, and D. Watts. 1970. "Forecasting Daily Hog Prices and Quantities: A Study of Alternative Forecasting Techniques." *Journal of American Statistical Association* 65:90–107.

Moschini, G. 1988. "A Model of Production with Supply Management for the Canadian Agricultural Sector." *American Journal of Agricultural Economics* 70:318–329.

Rausser, G. C. and D. P. Stonehouse. 1978. "Public Intervention and Producer Supply

Response." *American Journal of Agricultural Economics* 60(5):885–890.

Stonehouse, D. P. 1987. "A Profile of the Canadian Dairy Industry and Government Policies." Working paper 4/87. Ottawa, Ontario: Agriculture Canada, Policy Branch.

Stonehouse, D. P., D. H. Harrington, and R. K. Sahi. 1978. "An Econometric Forecasting and Policy Analysis Model of the Canadian Dairy Industry." In *Commodity Forecasting Models for Canadian Agriculture*, Vol. 1, edited by H. B. Huff and Z. A. Hassan. Ottawa, Ontario: Agriculture Canada, Economics Branch.

Stonehouse, D. P. and E. Kizito. 1990. "A Quarterly Econometric Model of the Canadian Dairy Industry." Working paper 4/90. Ottawa, Ontario: Agriculture Canada, Policy Branch.

Stonehouse, D. P. and G. C. Rausser. 1981. "A Quarterly Econometric Model for the Canadian Dairy Industry with Endogenous Policy Variables." Working paper AEEE 81/7. Guelph, Ontario: School of Agricultural Economics and Extension Education, University of Guelph.

Tinbergen, J. 1955. *On the Theory of Economic Policy*. Amsterdam: North Holland.

Veeman, M. M. 1982. "Social Costs of Supply-Restricting Boards." *Canadian Journal of Agricultural Economics* 30(1):21–36.

83-102

CHAPTER 6

Advertising, Product Promotion, and Market Demand

Q11 Q13

M37 M31

Hui-Shung Chang, Richard D. Green, and James Blaylock

THE DAIRY INDUSTRY is characterized by price-discriminating marketing orders, cooperatives, and price support programs. Furthermore, the existence of large dairy cooperatives, milk handlers, distributors, and processors indicates a relatively concentrated marketing channel for milk and dairy products (Kinnucan and Forker 1987). In such a complex industry, why do dairy farmers advertise and promote their products? The most basic explanation is provided by Gardner (1981). "Farmers as a group are better off at a given increase in the price of a product, if the increase comes about through an increase in demand rather than a decrease in supply." Since advertising and promotion of dairy products attempts to shift the demand curves outward, dairy farmers, from a welfare viewpoint, are better off than if the price increase were to come about from an inward shift of the supply function due to a particular production control program. If this is the case, then why do dairy producers not always prefer advertising and promotion techniques over other types of supply programs? The answer is simple. It is difficult to shift the total demand for food outward and, in particular, it is not easy to shift the total demand for dairy products outward enough to substantially improve the welfare of dairy producers.

How effective is advertising and promotion of dairy products? Empirical studies imply mixed results. Kinnucan (1987) found that Buffalo, New York, consumers were responsive in their fluid milk consumption to increases in advertising. He attributed part of this responsiveness to the overlapping media coverage with Ontario, Canada. Thompson and Eiler (1975) found that increases in profits of dairy producers from advertising vary substantially among markets. Blaylock et al. (1986) were not able to find significant effects of advertising on the demands for natural and processed cheese. There also exists the theoretical problem of the effects of imperfect competition on the returns of advertising to dairy producers. That is, if there is a high degree of concentration in the marketing channels for dairy products, then are middlemen the primary beneficiaries of increases in generic advertising of dairy products? To date, there has not been a study addressing the effectiveness of generic and brand advertising on the various stages of the marketing channel for dairy products.

There have been many changes in the dairy industry in recent years. For example, per capita milk consumption has decreased by approximately 20 percent since 1967 while per capita soft drinks consumption has more than doubled over the same time period, per capita consumption of cheese has increased by over 200 percent since 1967, and there has been a dramatic shift from whole milk to lowfat milk (Tauber and Forker 1987). To capture the interrelationships among the various commodities and the effects of advertising on these changes, it is more appropriate to model the demand for these commodities using demand systems or subsystems rather than employing single equation approaches.

The primary purposes of this chapter are to discuss some important theoretical issues related to advertising and promotion of dairy products and to report some empirical results from single equation and demand systems incorporating the effects of advertising on dairy products. The next section discusses the theoretical approaches or rationalizations for including advertising in demand estimation. This section concentrates on information, consumer processing of information, attitudes, and their relationships to estimable forms of demand functions. The third section of the chapter addresses the demand models incorporating advertising effects for dairy products. The empirical results from these models are reported next, and finally, the chapter presents conclusions and policy recommendations based on the empirical advertising/demand models.

THEORETICAL APPROACHES

There are a number of theories about the effects of advertising. A brief discussion of some of these theories is given here. It includes a marketing approach, an economic approach, and an information approach. This is a general review; however, in the empirical demand models these concepts are applied to dairy products.

THE MARKETING APPROACH

First, consider how marketing researchers view the role of advertising.

The theory of hierarchical effects is used by Lavidge and Steiner (1961). They postulate that consumers go through a series of psychological steps prior to purchase: awareness, interest, desire, and action. Advertising works by moving people more rapidly along these paths. According to this theory the effectiveness of advertising is measured by awareness, recall (attention), comprehension, and so on. The rationale behind this view is that advertising's ultimate function is to stimulate sales. However, an immediate sale response is not possible, that is, potential consumers seldom become customers as soon as they hear about the product. Rather, they approach the ultimate purchase in a series of steps.

Advertising then is a force that moves people up the series of steps by providing information and by persuasion. Thus, much of the effect of advertising is long-term in the sense that advertising is not, should not, and cannot be designed to produce immediate purchases. However, if something is to happen in the long run, something must first happen in the short run—something that will eventually lead to sales. Therefore, measurements of awareness, recall, and intention to buy can be used as predictive measurements for the ultimate influence of advertising on sales. Consumers are assumed to start with a complete lack of awareness of the existence of the product or service in question. When consumers see an advertisement, they first become aware of the product's existence and find out what it has to offer; next, they begin to like the product; then, to prefer the product over other alternatives. Gradually, they become convinced that the purchase would be wise and gain the desire to buy it. Finally, they make the actual purchase.

Palda (1966) casts doubt on the underlying assumption that progressing through the Lavidge-Steiner hierarchy of steps actually increases the probability of consumer purchase. Hence, measurements obtained on these "intermediate variables," which serve at best as predictors, may not account for the ultimate influence of advertising in increasing sales. Therefore, he suggests actual changes in sales as the correct measure for determining the effectiveness of advertising.

A second theory focuses on attitude changes (Achenbaum 1972). Achenbaum proposes that advertising works by helping consumers identify salient product attributes. Attribute comparisons, preferences, and attitude changes then are measurements for the effectiveness of advertising. Achenbaum disagrees with the view that consumers are ignorant, irrational, and at the mercy of the advertisers. Instead, he believes that consumers know what they want. Communication of information alone is not the only way consumer behavior is affected by advertising; rather, it is information about one or more product attributes salient to consumers that is communicated in a persuasive context by advertising. These attributes can be sensory, subjective, or emotional. If the communication is persuasive enough to improve consumers' attitudes about certain attributes, their attitudes toward the overall brand will improve as well, which, in turn, will increase the probability of purchase. Approaching consumers this way is not to be seen as manipulative; rather, it is that advertisers learn about what consumers want and cater to their wants. Thus, advertising is basically demand oriented.

Another view on the role of advertising advanced by Ehrenberg (1974) is that advertising works by reinforcing feelings of satisfaction for brands already being used. Ehrenberg claims that advertising is often effective but is not as powerful as is sometimes thought; he feels there is little evidence that advertising actually works by a strong form of persuasion or manipulation. A repeat buying habit may be developed and reinforced if there is satisfaction after previous usage. Therefore, for frequently bought products for which repeat buying is the main determinant of sales volume, advertising's main role is to reinforce feelings of satisfaction with brands already bought. In this context the role of repetitive advertising of well-established brands is defensive, used mainly to reinforce already developed repeat buying habits.

There are also theories assuming various roles for advertising such as to remind, to solve problems, to create an image, or to establish a brand personality. In an editorial, Tauber (1982) suggests moving away from the "one-measure-fits-all" practice as hypothesized by the above theories. Rather, the role of advertising varies depending on the situation. For instance, the hierarchical effects hypothesis seems most appropriate when a buying decision is in the purchase of durables or new products because of the greater psychological and/or economic commitment involved, whereas the reinforcement theory may be more appropriate for frequently bought products. Sheth (1974) summarizes some of the specific impacts that advertising may have on consumer behavior. He lists four separate mechanisms through which those impacts may be achieved:

1. Precipitation encourages consumers to become buyers of a product or service and speeds up the buy-no-buy decision for a general class of product or service.
2. Persuasion encourages consumers to choose among alternatives within the same general product or service class.
3. Reinforcement rationalizes consumers' choice behavior to build loyalty among purchases of the advertised brands.
4. Reminder encourages consumers to become habitual users for a particular brand or for a general class.

Precipitation and reminder are more likely to increase total industry sales, while persuasion and reinforcement are more likely to increase or help maintain a specific firm's market share. If this is the case then generic advertising should stress precipitation and reminder while brand-specific advertising should emphasize persuasion and reinforcement.

The Economic Approach

One of the objectives of advertising is to increase sales by shifting demand, either by shifting the demand curves or changing their shapes. Advertising may make demand either more or less elastic. The question arises, under what conditions would advertising make demand more elastic or less elastic? Two schools of thought have emerged with respect to how economists describe the economic effects of advertising on prices and price elasticities (see, e.g., Albian and Farris 1981). One school views advertising as a form of persuasion that creates product differentiation, making consumers less responsive to price changes and giving firms more market power. Galbraith (1967), for example, holds to this view of advertising. The other school maintains that advertising provides additional information to consumers thereby making markets more competitive and consumers more sensitive to price changes. For an excellent recent discussion of this viewpoint, see Ekelund and Saurman (1988). They attempt to explain why market-oriented economies such as those of the United States and Canada rank first and third, respectively, in advertising expenditures as a percentage of Gross National Product. The crucial difference between these two views is that, in the former, advertising increases product differentiation making the demand curve less elastic and leading to higher prices; in the latter, advertising makes demand curves more elastic and leads to lower prices. This view is held by Fergerson (1974), Nelson (1974), Stigler (1961), and Telser (1964).

This dichotomy is based on the advertising's content being either persuasive or informative. However, most researchers feel that both roles of advertising are present, although one may dominate the other under certain circumstances. For example, persuasion may prevail when (1) facts are unfavorable to the seller, (2) it is difficult to measure the quality embodied in a product, or (3) products are homogeneous. Qualities of a brand that consumers can determine by inspection prior to purchase are referred to as "search qualities" and qualities that are not determined prior to purchase are called "experience qualities" (Nelson 1974). Incidentally, Nelson classifies dairy products as experience nondurables. More than 40 years ago Borden (1942) noted that "advertising is informative in content, persuasive in intent."

Albion and Farris (1981) reviewed empirical studies of the effects of advertising on the price elasticities of demand and found that some showed that advertising decreased price sensitivity, while others reported increased sensitivity. Albion and Farris also point out that a correlation between advertising and price elasticity is not sufficient to infer the directions of causality; furthermore, decreased price sensitivity is not necessarily translated into a higher price, nor does increased price sensitivity entail a lower price. Therefore, it is necessary to differentiate between factory price and retail price elasticities and to determine the direction of causality to make any conclusions.

Questions have also been raised about the relative effectiveness of advertising versus prices. In some situations relative prices seem to have a greater influence on sales than advertising (Tilley and Lee 1980). However, Camanor and Wilson (1974) contend that advertising is a more important determinant of consumer expenditure than are relative prices. Also, Lee (1983) found that advertising is more effective than a price reduction in increasing sales for Florida grapefruit juice. In reality, it is difficult to separate the effect of advertising from prices. Not only are advertising and prices used to signal quality, but they are also often used simultaneously as part of the marketing mix. For example, advertising may be used to announce a price reduction.

Not only are there two schools of thought about the economic impacts of advertising on the structure, conduct, and performance of the market, there are divergent views on the relationship between advertising and a consumer's utility function. For example, do consumers receive utility directly from advertising or from what is being achieved by advertising? Dixit and Norman (1978) suggest entering advertising directly into the utility function as a tastes shifter. Such treatment, however, is criticized for (1) treating advertising as a direct generator of utility when in fact it is not (Fisher and McGowan 1979) and for (2) offering no understanding of the role of advertising in the consumer's decision process (Kotowitz and Mathewson 1979). In response, Kotowitz and Mathewson consider rational, although not fully informed, consumers whose buying pattern depends on their prior perceptions about aspects of product quality or attributes. Advertising affects consumers' perceptions, not tastes, by supplying information about attributes salient to consumers. Moreover, the utility that consumers get from advertising is derived from their improved perceptions of product quality, which leads to better purchase decisions.

Stigler and Becker (1977) also contend that tastes are stable over time. Based on the household production function approach (preferences are ordered over characteristics instead of over goods), advertising affects household consumption technology as an information provider. In traditional demand theory households maximize a utility function of the goods and services bought in the marketplace, whereas in the household production function approach they maximize a utility function of objects of choice, called commodities, which consumers produce with market goods, their own time, their skills, their stocks of knowledge, and other inputs. Stated formally, a household seeks to maximize its utility given a utility function

$$U = U(z_l, \ldots, z_m) \tag{6.1}$$

subject to the household production functions

$$z_i = f_i(x_{1i}, x_{2i}, \ldots, x_{ni}, T, A, O), \quad i = 1, \ldots, m \tag{6.2}$$

and full income

$$I = Wt + y \tag{6.3}$$

where z_i is the ith commodity objects of choice entering the utility function, f_i is the production function for the ith commodity, x_{ji} is the quantity of the jth market good or service used in the production of the ith commodity, A is the advertising expenditure or some measure used as a proxy for information or stocks of knowledge, and O represents other factors that are relevant in the household production process; w is the wage rate, T are the hours worked, and y is the nonwage income.

Consumer behavior depends on prices, income, and tastes as well as the household consumption technology. Moreover, demand for market goods changes not because advertising changes tastes, but because the information provided by advertising reduces the costs or shadow prices of commodities produced by the household. Thus, the demand for their inputs (market goods) increases. The implications of the household production function approach are richer than the neoclassical approach to demand. First, the relative magnitudes of the cross-price elasticities of the market goods depend on their elasticity of substitution in production. That is, goods that can produce more similar characteristics are considered as substitutes. Second, the household production function approach justifies the use of the houschold as a basic unit of observation (Michael and Becker 1973). Most important, analyses no longer depend upon "changes in tastes" to explain phenomena such as the effects of advertising and addiction to smoking or music (Stigler and Becker 1977).

As attractive as the Stigler-Becker approach may appear, there are some operational problems. Some analysts believe that advertising affects consumer purchase decisions, not the production decisions after purchase. Second, application of the "new" approach requires strong assumptions about the household production technology. In particular, the two-stage optimization procedure requires technology to be linearly homogeneous and to have an absence of joint production. Violation of these conditions results in prices being endogenous, which may complicate the analysis substantially

(Deaton and Muellbauer 1980b). Furthermore, the inclusion of nonmarket variables in the analysis is bound to encounter more definitional and measurement problems than the traditional approach. Finally, an integration of the theory of the consumer and the theory of the firm may cause some identification problem in addition to requiring a labor supply function. Distinguishing goods from their characteristics, Green (1978) contends that preferences related to characteristics change less over time than those related to goods. Therefore, advertising may change people's preferences for a particular brand or product by providing information about its attributes or characteristics, but it does not change their preferences for characteristics.

THE INFORMATION APPROACH

A discussion of advertising would not be complete without consideration of risk and uncertainty. Definitions from information theory relate information to uncertainty and the statistical concept of variance. That is, uncertainty may be summarized by assigning a probability distribution over possible states of nature, and information may be defined as events that tend to change these distributions (Hirschliefer 1973). More frequently, the price and/or quality differential existing in the market are assumed to be uncertain to the consumers. Stigler (1961) and Nelson (1970) state that after receiving information, the decision maker is more certain of the ultimate result of the decision; thus, information reduces uncertainty. In other words, under uncertainty when consumers attempt to maximize their expected utility, their ex post expected utility with information is no less than their ex ante expected utility before the information is acquired. The difference between the two utility levels is then defined as the value of information, which is positive.

Some researchers, however, have maintained that information could actually increase uncertainty. Information is usually costly; it may either involve time cost or actual out-of-pocket costs; therefore, consumers have to weigh the costs against the benefits when engaging in information search. If advertising provides "free" information or reduces search cost, then there should be some informational value associated with advertising, assuming the information provided is not faulty or misleading. Also advertising is considered a "type of public good" in the sense that the service or information provided by advertising is accessible to everyone indiscriminately, irrespective of whether consumers buy the service or not. (Security, defense, and street lighting are other examples of public goods.) Therefore, the services can only be provided by taxation or subsidized by the "advertised" commodity and not by individual purchase and sale (Kaldor 1950).

Pope (1985) asserts that when risk and uncertainty (about product quality or characteristics) are involved, nothing can be said a priori, not even about the signs of advertising elasticities; that is, advertising may increase or decrease the demand for advertised products. In a situation of information overload, where too much information is being supplied, consumers may be more confused than before; therefore, decision making is not enhanced but rather hindered by additional information.

DEMAND MODELS INCORPORATING ADVERTISING EFFECTS

This section presents results from single equation models that are developed to determine the influence of advertising on natural and processed cheese. Emphasis is given to the potential carryover effects of advertising on these dairy products. In addition, aggregate food demand subsystems that explicitly include advertising effects are developed. The effects of advertising on dairy products (an aggregate food group) will be the primary focus of this presentation. Selected concepts discussed in the previous theoretical section are incorporated in these models.

SINGLE EQUATION MODELS

In single equation models, advertising enters the demand equation directly as another explanatory variable. The usual assumption is that demand for a product depends not only on its price and consumer's income, but also on how much advertising is provided.

Since the demand models for natural and processed cheese are reported in Blaylock et al. (1986), the focus of this discussion relates to the effects of advertising on the demands for these products. The demand model for natural cheese can be expressed as

$$\begin{aligned}\ln q_t^n = {} & \beta_0 + \beta_1 \ln(P_t^n) + \beta_2 \ln(P_t^p) + \beta_3 \ln(P_t^I) + \beta_4 \ln(P_t^m) + \\ & \beta_5 \ln(Y_t) + \beta_6 D_t + \beta_7 T_t + \sum_{i=1}^{12} \delta_i M_i + \\ & \sum_{j=0}^{12} \gamma_j GA_{t-j} + \Sigma \theta_j BA_{t-j} + \epsilon_t \end{aligned} \tag{6.4}$$

where

q_t^n: per capita quantity of natural cheeses purchased by households in the United States, in pounds, per month t, t = 1 (January 1982), . . . , t = 42 (June 1985)

P_t^n: price of natural cheese in dollars per pound, deflated by the

Consumer Price Index (CPI) for all urban consumers (1977 = 100)

P_t^p: price of processed cheese in dollars per pound, deflated by the CPI

P_t^I: price of imitation cheese in dollars per pound deflated by the CPI

P_t^m: price index of meat, poultry, and fish (1967 = 100) deflated by the CPI

Y_t: per capita disposable income in the United States is period t, deflated by the CPI

D_t: per capita domestic donations of cheese under the U.S. Department of Agriculture's Temporary Emergency Food Assistance Program, in pounds

T_t: time trend, $t = 1$ (January 1982), . . . , $t = 42$ (June 1985)

M_i: monthly dummy variables, $M_1 = 1$ if January, zero otherwise; $M_2 = 1$ if February, zero otherwise, etc. December is omitted to avoid perfect multicollinearity

$\sum_{j=0}^{12} \gamma_j GA_{t-j}$: weighted average of current and past per capita generic advertising expenditures on real cheese in period t. Generic advertising expenditures are "deflated" by a media cost index

$\sum_{j=0}^{12} \theta_j BA_{t-j}$: weighted average of current and past per capita branded advertising expenditures on real cheese in period t. Branded advertising expenditures are "deflated" by a media cost index

$\delta_i, \beta_i, \gamma_j, \theta_j$: model parameters to be estimated

ϵ_t: equation error term, which is assumed to be normally distributed with zero mean and constant variance.

The specification of the processed cheese equation is similar and will not be given here; see Blaylock et al. (1986) for details.

The empirical results of the demands for natural and processed cheese are given in Tables 6.1 and 6.2, respectively. The discussion will be restricted to advertising affects. First, both branded and generic advertising variables are lagged 12 time periods in order to capture the carryover effects (recall that the data consist of monthly observations). In the natural cheese equation, the signs on the advertising coefficients are "incorrect" (see Table 6.1) however, the signs of the advertising coefficients in the processed cheese equation conform to a priori reasoning (see Table 6.2). In

Table 6.1. Parameter estimates: Natural cheese equation

Variable	Coefficent	Standard error	Level of significance
			2%
Intercept	−9.882	5.745	10
Price of natural cheese	−1.531	0.767	6
Price of processed cheese	0.653	0.528	23
Price of imitation cheese	−0.143	0.148	35
Price index for meat, poultry, and fish	0.409	0.360	27
Disposable personal income	1.272	0.718	9
Government donations	−0.013	0.013	35
M1(= 1 if January, zero otherwise)	−0.135	0.037	1
M2(= 1 if February, zero otherwise)	−0.252	0.061	1
M3(= 1 if March, zero otherwise)	−0.136	0.033	1
M4(= 1 if April, zero otherwise)	−0.207	0.025	1
M5(= 1 if May, zero otherwise)	−0.209	0.035	1
M6(= 1 if June, zero otherwise)	−0.255	0.029	1
M7(= 1 if July, zero otherwise)	−0.271	0.040	1
M8(= 1 if August, zero otherwise)	−0.233	0.044	1
M9(= 1 if September, zero otherwise)	−0.233	0.035	1
M10(= 1 if October, zero otherwise)	−0.158	0.040	1
M11(= 1 if November, zero otherwise)	−0.151	0.024	1
Trend(= 1 if Jan., 1982, 2 if Feb., 1982, etc.)	−0.009	0.003	1
Branded advertising, current period	−0.013	0.030	67
Branded advertising, lagged 1 period	−0.024	0.056	67
Branded advertising, lagged 2 periods	−0.033	0.077	67
Branded advertising, lagged 3 periods	−0.040	0.093	67
Branded advertising, lagged 4 periods	−0.045	0.105	67
Branded advertising, lagged 5 periods	−0.048	0.112	67
Branded advertising, lagged 6 periods	−0.049	0.114	67
Branded advertising, lagged 7 periods	−0.048	0.112	67
Branded advertising, lagged 8 periods	−0.045	0.105	67
Branded advertising, lagged 9 periods	−0.040	0.093	67
Branded advertising, lagged 10 periods	−0.033	0.077	67
Branded advertising, lagged 11 periods	−0.024	0.056	67
Branded advertising, lagged 12 periods	−0.013	0.030	67
Generic advertising, current period	−0.116E-5	0.738E-4	13
Generic advertising, lagged 1 periods	−0.214E-5	0.136E-4	13
Generic advertising, lagged 2 periods	−0.294E-5	0.187E-4	13
Generic advertising, lagged 3 periods	−0.356E-5	0.227E-4	13
Generic advertising, lagged 4 periods	−0.401E.5	0.255E-4	13
Generic advertising, lagged 5 periods	−0.427E-5	0.272E-4	13
Generic advertising, lagged 6 periods	−0.436E-5	0.278E-4	13
Generic advertising, lagged 7 periods	−0.427E-5	0.272E-4	13
Generic advertising, lagged 8 periods	−0.401E-5	0.255E-4	13
Generic advertising, lagged 9 periods	−0.356E-5	0.227E-4	13
Generic advertising, lagged 10 periods	−0.294E-5	0.187E-4	13
Generic advertising, lagged 11 periods	−0.214E-5	0.136E-4	13
Generic advertising, lagged 12 periods	−0.116E-5	0.738E-4	13

Summary statistics: Adjusted $R^2 = 0.96$; Durbin-Watson = 1.92.

Source: Blaylock et al. (1986).

Table 6.2. Parameter estimates: Processed cheese equation

Variable	Coefficent	Standard error	Level of significance
			2%
Intercept	-1.660	1.382	24
Price of natural cheese	0.749	0.423	9
Price of processed cheese	−0.089	0.675	9
Price of imitation cheese	0.216	0.215	32
Price index for meat, poultry, and fish	0.157	0.471	33
Disposable personal income	0.166	0.178	94
Government donations	−0.065	0.014	1
Dummy (= 1 if June, July, or August)	−0.021	0.014	14
Branded advertising, current period	0.767E-6	0.670E-6	26
Branded advertising, lagged 1 period	−0.142E-5	0.124E-5	26
Branded advertising, lagged 2 periods	0.195E-5	0.170E-5	26
Branded advertising, lagged 3 periods	0.236E-5	0.206E-5	26
Branded advertising, lagged 4 periods	0.266E-5	0.232E-5	26
Branded advertising, lagged 5 periods	0.283E-5	0.247E-5	26
Branded advertising, lagged 6 periods	0.289E-5	0.253E-5	26
Branded advertising, lagged 7 periods	0.283E-5	0.247E-5	26
Branded advertising, lagged 8 periods	0.266E-5	0.232E-5	26
Branded advertising, lagged 9 periods	0.236E-5	0.206E-5	26
Branded advertising, lagged 10 periods	0.195E-5	0.170E-5	26
Branded advertising, lagged 11 periods	0.142E-5	0.124E-5	26
Branded advertising, lagged 12 periods	0.767E-5	0.670E-5	26
Generic advertising, current period	0.037	0.031	25
Generic advertising, lagged 1 periods	0.068	0.058	25
Generic advertising, lagged 2 periods	0.094	0.080	25
Generic advertising, lagged 3 periods	0.114	0.096	25
Generic advertising, lagged 4 periods	0.128	0.109	25
Generic advertising, lagged 5 periods	0.137	0.109	25
Generic advertising, lagged 6 periods	0.139	0.118	25
Generic advertising, lagged 7 periods	0.137	0.116	25
Generic advertising, lagged 8 periods	0.128	0.109	25
Generic advertising, lagged 9 periods	0.114	0.096	25
Generic advertising, lagged 10 periods	0.094	0.080	25
Generic advertising, lagged 11 periods	0.068	0.058	25
Generic advertising, lagged 12 periods	0.037	0.031	25

Summary statistics: Adjusted $R^2 = 0.83$; Durbin-Watson = 1.86.

Source: Blaylock et al. (1986).

both of the demand functions all of the advertising effects are not statistically different from zero at a reasonable level of significance. Thus, individual advertising effects, current and lagged, do not significantly influence the quantities demanded of either natural or processed cheese. However, the high adjusted R^2 values together with the large number of insignificant coefficients may indicate a high degree of multicollinearity.

In order to test whether or not either branded or generic advertising collectively has a significant impact on the quantities demanded of natural or processed cheese, partial F tests need to be performed. Blaylock et al. (1986) did not report the results of these tests. Hence, even though Blaylock et al. were not able to detect significant individual effects of advertising on dairy products, aggregate or collective advertising effects may still exist.

Single equation models do not account for the interrelationships among commodities. Hence, a demand system explicitly incorporating the effects of advertising is presented in the next subsection. Since there are severe data limitations associated with the demand system estimations, the emphasis will be placed on the theoretical or conceptual frameworks for analyzing the effects of advertising on food commodities.

Demand Systems

The questions to be answered, or hypotheses to be tested, using a demand system approach are: Does advertising change consumers' buying behavior? Does advertising make consumers more responsive or less responsive to price and income changes? And does advertising for one commodity affect the demands for other commodities? To address these questions, the almost ideal demand system (AIDS) of Deaton and Muellbauer (1980a, 1980b), which has several desirable theoretical properties, is used. Advertising effects are explicitly incorporated into the demand system using the translation method of Pollak and Wales (1980). Classical static demand theory assumes perfect information and constant tastes. However, consumers frequently possess less than perfect information. Moreover, tastes may change due to better information or changes in consumers' environments such as their sociodemographic situation. Tintner (1951) and Ichimura (1950–51) defined changes in tastes as an alteration in the form of the ordinal utility function or indifference map. In this case, parameters of the utility functions are postulated to depend on consumers' states of knowledge and some economic or noneconomic variables.

The almost ideal demand system in share form including advertising effects is given by

$$\begin{aligned} w_{it} = {} & (a_i + b_i A_{i,t-1}) + \Sigma \gamma_{ij} \ln p_{j,t} \\ & + \beta_i (\ln y_t - \alpha_0 - \Sigma (a_j + b_j A_{j,t-1}) \ln p_{j,t} \\ & - 1/2 \Sigma \Sigma \gamma_{ij} \ln p_{i,t} \ln p_{j,t}) \end{aligned} \qquad (6.5)$$

where w_{it} represents the ith budget share, A_i is the advertising expenditures on commodity i, P_i are prices, and y_t is total expenditures. In addition, $\Sigma\beta_i = 0$, $\Sigma \gamma_{ij} = 0$, and $\Sigma (a_j + b_j A_{j,t-1}) = 1$. Each γ_{ij} represents the change in the ith budget share with respect to a percentage change in the jth, price, with real expenditure held constant. Each β_i represents the change in the ith budget share with respect to a percentage change in the real income or expenditures with prices held constant.

Advertising effects are introduced, following Green (1985), by allowing some of the parameters in equation (6.5) to depend on past advertising expenditures. For a theoretical justification for this approach, see Blanciforti et al. (1985). Based on equation (6.5) expressions for the expenditure, own-price, cross-price, and advertising elasticities are given, respectively, by expressions in (6.6), (6.7), (6.8), (6.9), and (6.10). Given the limited data set, A_{t-1} will simply be the previous own-advertising expenditures.

$$\eta_{it} = 1 + \beta_i / w_{i,t} \tag{6.6}$$

$$\begin{aligned} \epsilon_{ii,t} = -1 + \{\gamma_{ii} - \beta_i[(a_i + b_i A_{i,t-1}) \\ + \Sigma \gamma_{ij} \ln p_{j,t}]\} / w_{i,t} \end{aligned} \tag{6.7}$$

$$\begin{aligned} \epsilon_{ij,t} = \{\gamma_{ij} - \beta_i[(a_j + b_j A_{j,t-1}) \\ + \Sigma \gamma_{ij} \ln p_{j,t}]\} / w_{i,t} \end{aligned} \tag{6.8}$$

$$\epsilon_{iAi,t} = [b_i (1 - \beta_i \ln p_{i,t}) A_{i,t-1}] / w_{i,t} \tag{6.9}$$

$$\epsilon_{iAj,t} = -(b_j \beta_i \ln p_{j,t}) A_{j,t-1} / w_{i,t} \tag{6.10}$$

Consider the own-advertising elasticity, $\epsilon_{iAi,t}$, given in equation (6.9). It represents the percentage change in the quantity demanded of the ith commodity with respect to a percentage change in the expenditures of advertising devoted to commodity i in the previous period. A priori the own-advertising elasticities are expected to be nonnegative; otherwise, additional amounts of advertising directed toward a particular commodity would result in a reduction of the quantities demanded. For necessities ($0 < \eta < 1$) whose income elasticities are positive but less than one, the AIDS implies $\beta_i < 0$ [refer to equation (6.6)]. Consequently, with $b_i > 0$, the own-advertising elasticity is always positive, assuming proper scaling on prices so that $\ln p_{i,t}$ is positive. However, for luxuries ($\eta > 1$) the AIDS implies that $\beta_i > 0$. Hence, with $b_i > 0$ the advertising elasticity is positive only if $(1 - \beta_i \ln P_{i,t}) > 0$, assuming proper scaling on prices. The cross-advertising elasticities also depend upon the signs and the relative magnitudes of the β_i and b_j, where b_j is the advertising coefficient for commodity j.

Now, consider the effects of advertising on the elasticities of demand.

First, with respect to the expenditure elasticities

$$\partial \eta_{i,t} / \partial A_{i,t-1} = -[\beta_i b_i (1 - \beta_I \ln p_{i,t})] / w_{i,t}^2. \qquad (6.11\}$$

Thus, for necessities ($\beta_i < 0$) an increase in the advertising expenditure of the ith commodity results in the expenditure elasticity becoming more elastic or less inelastic, assuming that $b_i > 0$ and proper scaling of prices. Second, with respect to own-price elasticities, the effect of advertising is

$$\partial \epsilon_{ii,t} / \partial A_{i,t-1} = -(\beta_i b_i / w_{i,t}) - \{-1 + \{\gamma_{ii} - \beta_i [(a_i + b_i A_{i,t-1}) + \Sigma \gamma_{ij} \ln p_{j,t}]\}\} \{b_i (1 - \beta_i \ln p_{i,t})\} / w_{i,t}^2. \qquad (6.12)$$

Third, with respect to cross-price elasticities, the effect of advertising is

$$\partial \epsilon_{ij,t} / \partial A_{i,t-1} = -\{\gamma_{ij} - \beta_i [(a_j + b_j A_{j,t-1}) + \Sigma \gamma_{ij} \ln p_{j,t}]\} \{b_i (1 - \beta_i \ln p_{i,t})\} / w_{i,t}^2. \qquad (6.13)$$

Finally, with respect to own-advertising elasticities, the effect of advertising is

$$\partial \epsilon_{iA,t} / \partial A_{i,t-1} = [b_i (1 - \beta_i \ln p_{i,t})] + [w_{i,t} - b_i (\beta_i \ln p_{i,t}) A_{i,t-1}] / w_{i,t}^2. \qquad (6.14)$$

Clearly, the complex expressions of equations (6.11), (6.12), (6.13), and (6.14) make it difficult to determine the effects of advertising on these elasticities a priori. Therefore, there exists sufficient flexibility in the model to allow these important questions to be answered empirically.

The model was estimated for five food groups: meats, dairy products, cereal and bakery products, fruits and vegetables, and all other foods consumed at home. The dairy products group includes milk, butter, eggs, cheese, and ice cream.

Advertising expenditures were obtained from BAR/LNA multimedia service data series collected by Broadcast Advertisers Reports, Inc., and Leading National Advertisers, Inc. The series contains quarterly advertising expenditures on various products by class, item, and company from 1980 to 1984. Advertising expenditures for the dairy products are then aggregated according to the above classification. The aggregate advertising expenditures generated this way include generic advertising from various marketing programs. The advertising expenditures are deflated by a media deflator [see Chang (1989) for details].

Full information maximum likelihood estimates methods were employed to obtain coefficient estimates using the computer algorithm SHAZAM (White 1978). Because of data limitations, it was not possible to estimate the complete AIDS. Instead, a linear approximation to this system suggested by Deaton and Muellbauer (1980a) was employed. In the linear approximate AIDS (LA/AIDS) the Stones's index, $\ln P^* = \Sigma w_k \ln p_k$, was used to replace the price index $\ln P$ given by the last term in parenthesis in equation (6.5). Thus, the AIDS system actually estimated is given by

$$w_{it} = a_i^* + \Sigma \gamma_{ij} \ln p_{j,t} + \beta_i \ln (y_t/P_t^*) \quad (6.15)$$

where a_i^* is assumed to be a function of previous advertising expenditures. Some estimation problems associated with this approximation are discussed in Blanciforti and Green (1983).

For empirical purposes an additive error structure is assumed for the LA/AIDS in equation (6.15). In addition, autocorrelated error terms are assumed, $e_{i,t} = \rho_i e_{i,t-1} + u_{i,t}$, where $\phi u_t \sim N(0, \sigma^2 I_T)$ and ρ_i is the same for all equations.

Since the budget shares sum to one, it follows that the contemporaneous covariance matrices are singular. In the absence of autocorrelation, Barten (1969) has shown the estimates of the parameters can be obtained from full information maximum likelihood methods by arbitrarily deleting one equation. The resultant estimates are invariant to the choice of the equation deleted. When autocorrelation is present, standard corrective procedures are used to purge the equations of autocorrelation. The invariance property still holds as long as the coefficients of correlation, ρ_i, are identical for all equations (Berndt and Savin 1975).

The results for the linear approximation of the AIDS are presented in Tables 6.3 and 6.4. Due to space limitations only the estimated elasticities and the effects of advertising on estimated elasticities are presented. The expenditure elasticity estimate indicates that the meat group is a relative luxury (η = 1.660). For all other food groups the expenditure elasticity estimates were less than one, indicating that these commodities are relative necessities. With respect to the own-price elasticities, a value of −0.995 was obtained for meats and a value of −0.123 for dairy products (see Table 6.3, column 2).

All of the own-advertising elasticities estimates were positive except for cereal and bakery products (see Table 6.3). One possible explanation for this exception is that this group has a relatively small percentage of advertising that can be considered as generic advertising, that is, advertising that is done by either producer cooperatives or under the provisions of the industry marketing programs with market expansion as the major goal.

Table 6.3. The dynamic linear approximate almost ideal demand system, calculated own-elasticities (with autocorrelation, $\rho = 0.15$)

Food groups	Calculated elasticities[a]		
	η_i	ϵ_{ii}	$\epsilon_{i,Ai}$
Meats	1.660	−0.995	0.011
Dairy products	0.861	−0.123	0.030
Cereal and bakery products	0.667	−1.242	−0.006
Fruits and vegetables	0.808	0.183	0.029
All other foods at home	0.598	1.243	—[b]

[a] Elasticities are evaluated at sample means.
[b] Since the *b* for all other foods at home is not identifiable, the corresponding elasticity is not available.

Table 6.4. The dynamic linear approximate almost ideal demand system, effects of advertising on calculated elasticities (with autocorrelation, $\rho = 0.15$)

Foods groups	Effects of advertising on calculated elasticities[a]		
	$\partial\eta_i/\partial A_{i,t-1}$	$\partial\epsilon_{ii}/\partial A_{i,t-1}$	$\partial\epsilon_{i,A}/\partial A_{i,t-1}$
Meat	−0.005	0.038	0.008
Dairy products	0.003	0.137	0.020
Cereal and bakery products	−0.0003	−0.008	0.017
All other foods at home	—[b]	—	—

[a] The effects of advertising on elasticities are evaluated at sample means.
[b] The figures are not applicable due to nonidentifiability.

Generic advertising accounted for roughly 50 percent of total advertising expenditures for dairy products, about 26 percent for meats, and about 6 percent for fruits and vegetables in 1985. As for cereals and bakery products, even though they are heavily advertised much of the advertising is designed to protect market shares as opposed to increasing total industry sales.

The effects of advertising on elasticity values are presented in Table 6.4. Increases in advertising expenditures have negative effects on the expenditure elasticities for meats and cereals and bakery products (the estimates are −0.006 for meats and −0.0003 for cereals and bakery products). (See the first column of Table 6.4.) An increase in advertising expenditures increases the expenditure elasticities for dairy products and fruits and vegetables.

In all cases, the effect of advertising on own-price and own-advertising elasticities is to make the former less negative and the latter more positive for all the groups with the exception of cereal and bakery products (see columns two and three of Table 6.4). A positive effect of advertising on own-advertising elasticity may reflect the cumulative effect of advertising.

Does advertising create an "attachment" to products and make consumers less price responsive? Based on the AIDS, this appears to be the case; however, it is difficult to make strong inferences in this study since the commodities are so broadly defined and the data base was extremely limited; however, these are interesting ideas for further examination. Metwally (1980) in a study of eight Australian products found that promotional efforts of the sellers of the products seem to have succeeded in creating a significant degree of consumer "loyalty." Thus, there is some empirical evidence from other countries supporting the results obtained here.

CONCLUSION

This study indicates that it is possible to explicitly incorporate advertising effects into complete demand systems and obtain estimates of these effects on expenditure and price elasticities. Although the empirical results were based on a severely limited data base, they do indicate some potentially interesting and useful results. In the LA/AIDS, own-advertising effects appear to be important for dairy products; although not reported here, cross-advertising effects appear to be important for meats; advertising has positive effects on own-price elasticities for meats, dairy products, and fruits and vegetables.

There are also trade-offs between choosing a system approach and a single equation model. The former is theoretically plausible and allows the researcher to analyze the interrelationships among commodities. However, a system approach has less flexibility in the choice of the functional form and may impose restrictions that are not otherwise necessary. The single equation models for cheese products illustrate this.

The primary contribution of this chapter has been to demonstrate that it is possible to explicitly incorporate advertising effects into food demand subsystems and determine their effects on price, income, and advertising elasticities. However, due to the restrictive data base, the empirical results should be interpreted with caution. This research currently is being extended by the use of a combined cross-sectional and time-series data base. These results will allow more precise estimates upon which to base policy recommendations.

REFERENCES

Achenbaum, A. S. 1972. "Advertising Doesn't Manipulate People." *Journal of Advertising Research*, April, 3–13.

Albion, M. and P. Farris. 1981. *The Advertising Controversy: Evidence on the Economic Effects of Advertising*. Boston: Auburn House.

Barten, A. P. 1969. "Maximum Likelihood Estimation of a Complete System of Demand Equations." *European Economic Review* 1:7–73.

Berndt, E. R. and N. E. Savin. 1975. "Estimation and Hypothesis Testing in Singular Equation Systems with Autoregressive Disturbances." *Econometrica* 43:937–957.

Blanciforti, L. and R. Green. 1983. "An Almost Ideal Demand System Incorporating Habits: An Analysis of Expenditures on Food and Aggregate Commodity Groups." *Review of Economics and Statistics* 65:511–515.

Blanciforti, L., R. Green, and G. King. 1985. "U.S. Consumer Behavior Over the Postwar Period: An Almost Ideal Demand System Analysis." Giannini Foundation Monograph No. 40, Berkeley, University of California.

Blaylock J., D. Smallwood, and L. Myers. 1986. "Econometric Analysis of the Effects of Advertising on the Demand for Cheese." In *Report to Congress on the Dairy Promotion Program*. Washington, D.C.: ERS, USDA.

Borden, N. H. 1942. *The Economic Effects of Advertising*. Chicago, Ill.: Irwin.

Camanor, W. and T. Wilson. 1974 "The Effect of Advertising on Competition: A Survey." *Journal of Economic Literature* 17:453–476.

Chang, H. S. 1987. "Measuring the Effects of Advertising in Food Demand Subsystems." Ph.D. diss., University of California, Davis.

Deaton, A. and J. Muellbauer. 1980a. "An Almost Ideal Demand System." *American Economic Review* 70:312–326.

_____. 1980b. Economics and *Consumer Behavior*. Cambridge: Cambridge University Press.

Dixit, A. and V. Norman. 1978. "Advertising and Welfare." *The Bell Journal of Economics* 10:1–17.

Ehrenberg, A. S. 1974. "Repetitive Advertising and the Consumer." *Journal of Advertising Research*, April, 25-34.

Ekelund, Jr., R. and D. Saurman, 1988. *Advertising and the Market Process: A Modern Economic View*. San Francisco, Calif.: Pacific Research Institute for Public Policy.

Fergerson, J. 1974. *Advertising and Competition: Theory, Measurement, Fact*. Cambridge, Mass.: Ballinger.

Fisher, F. and J. McGowan. 1979. "Advertising and Welfare: Comment." *The Bell Journal of Economics* 11:726–727.

Galbraith, J. K. 1967. "The New Industrial State." Boston: Houghton Mifflin.

Gardner, B. 1981. The *Governing of Agriculture*. Lawrence, Kans.: University Press of Kansas.

Green, H. A. J. 1978. *Consumer Theory*. New York: Academic Press.

Green, R. 1985. "Dynamic Utility Functions for Measuring Advertising Responses." In *Proceedings of Research on Effectiveness of Agricultural Commodity Promotion Seminar*, 80–88. Sponsored by the Farm Foundation and the U.S. Department of Agriculture.

Hirschliefer, J. 1973. "Where Are We in the Theory of Information." *American Economic Review* 63:31–39.

Ichimura, S. 1950–51. "A Critical Note on the Definition of Related Goods." *Review of Economic Studies* 18:179–183.

Kaldor, N. 1950. "The Economic Aspects of Advertising." *Review of Economic Studies* 18(1):1–27.

Kinnucan, H. 1987. "Effect of Canadian Advertising on Milk Demand: The Case of the Buffalo, New York, Market." *Canadian Journal of Agricultural Economics* 35:181–196.

Kinnucan, H. and O. Forker. 1987. "Asymmetry in Farm-Retail Price Transmission for Major

Dairy Products." *American Journal of Agricultural Economics* 69:285–292.

Kotowitz, Y. and F. Mathewson. 1979. "Advertising, Consumer Information, and Product Quality." *The Bell Journal of Economics* 10:566–588.

Lavidge, R. C. and G. Steiner. 1961. "A Model for Predictive Measurements of Advertising Effectiveness." *Journal of Marketing* 25:59–62.

Lee, J. Y. 1983. "Florida Department of Citrus Advertising Research Program." In *Advertising and the Food System*, ed. J. M. Connor and R. W. Ward. North Central Regional Project NC-117, Mono. 14. Madison, Wis.: University of Wisconsin.

Michael, R. T. and G. S. Becker. 1973. "On the New Theory of Consumer Behavior." *Swedish Journal of Economics*: 75:378–396.

Nelson, P. 1970. "Information and Consumer Behavior." *Journal of Political Economy* 78:311–329.

_____. 1974. "Advertising as Information." *Journal of Political Economy* 82:729–754.

Palda, K. S. 1966. "The Hypothesis of a Hierarchy of Effects: A Partial Evaluation." *Journal of Marketing Research* 3:13–24.

Pollak, R. A. and T. J. Wales. 1980. "Comparison of the Quadratic Expenditure System and Translog Demand Systems with Alternative Specifications of Demographic Effects." *Econometrica* 48:595–612.

Pope, R. 1985. "The Impact of Information on Consumer Preferences." In *Proceeding of Research on Effectiveness of Agricultural Commodity Promotion Seminar*, 69–79. Sponsored by Farm Foundation and the U.S. Department of Agriculture.

Sheth, J. N. 1974. *Models of Buyer Behavior: Conceptual, Quantitative, and Empirical.* New York: Harper and Row.

Stigler, G. 1961. "The Economics of Information." *Journal of Political Economy* 69:213–225.

Stigler, G. and G. Becker. 1977. "De Gustibus Non Est Disputandum." *American Economic Review* 67:76–90.

Tauber, E. M. 1982. "Editorial: How Does Advertising Work?" *Journal of Advertising Research* 22:7.

Tauber, E. M. and O. Forker. 1987. "Dairy Promotion in the U.S.: 1979–1986." Ithaca: Department of Agricultural Economics, Cornell University.

Telser, L. 1964. "Advertising and Competition." *Journal of Political Economy* 72:537–562.

Thompson, S. and D. Eiler. 1975. "Producer Returns from Increased Milk Advertising." *American Journal of Agricultural Economics* 57:505–508.

Tilley, D. S. and J. Lee. 1980. "Import and Retail Demand for Orange Juice in Canada." ERS Report 80-1. Gainesville: Economic Research Department, Florida Department of Citrus.

Tintner, G. 1952. "Complementarity and Shifts in Demand." *Metroeconomica* 4:1–4

White, K. 1978. "A General Computer Program for Econometric Models—SHAZAM." *Econometrica* 46:239–240.

CHAPTER 7

A General Conceptual Framework for the Analysis of Advertising Impacts on Commodity Demand

Thomas L. Cox

CHAPTERS 4 AND 6 of this book summarize the relevant empirical analytics for the analysis of dairy product demand. In particular, Chapter 4 presents the empirical demand systems viewpoint while Chapter 6 summarizes the "classical" theoretical approach to analyzing the effects of advertising on commodity demand. This chapter presents an additional overview of the methodology for modeling the impacts of advertising on commodity demand from a general conceptual perspective.

In essence, the influence of advertising on consumption can be viewed as an additional ceteris paribus factor relevant to the estimation of commodity demand functions. However, quite in contrast to consumer demographics and disposable income, advertising is likely to be a factor amenable to some degree of control by commodity groups and/or firms. Hence, while accurate estimates of all major demand shifters are of general research interest, estimation of the impacts of advertising on consumption has generated a large and growing research literature.

As summarized in Chapter 6, the hypothesis that advertising somehow shifts commodity demand curves is common across various theoretical approaches found in the advertising/sales response literature. There is lively debate concerning how this shift occurs and, correspondingly, how advertising impacts should be modeled. Much of the business/marketing literature concerns firm level, brand-specific advertising. In contrast, much of the agricultural economics literature considers industry level, generic advertising. Some of the literature is concerned solely with accurate prediction of the impacts of advertising and the derivation of "optimal" advertising strategies from a firm (brand) or industry (generic) level perspective. In contrast, there is a considerable literature on consumer and societal welfare implications of advertising. Research objectives and the resultant conceptual frameworks appropriate for this latter context can be quite different from the former.

A key issue for all advertising and promotion research concerns how one conceptualizes and specifies the impacts of advertising on demand. This chapter assumes that advertising can be insightfully analyzed as an explicit shifter of consumer preferences. While this premise is an open debate in the literature (e.g., Kotowitz and Mathewson 1979), it is argued that the analytical benefits of the resulting conceptual framework outweigh the costs of a priori structure imposed on a complex measurement problem. Within this context the impacts of advertising on demand are analyzed under a generalized augmentation hypothesis following the work of Hanemann (1981). This conceptual framework is a specialization of the household production model. The recent work of Amuah (1985), Green (1985), Ward and Chang (1985), and the work of Ellen Goddard in Chapter 13 of this book can be viewed as applications of this generalized augmentation hypothesis.

This conceptual viewpoint suggests that much of the analysis of technical change in production literature and associated insight is relevant to the analysis of the advertising/consumption response. If this observation is correct, then several serious identification and specification issues common to the analysis of structural change confront the analysis of advertising impacts on consumption. The identification issues are discussed in the next section. Within this context a general conceptual framework that addresses these issues from a household production viewpoint is then developed. Following this discussion four specification alternatives as special cases of the general framework are identified. Some caveats on the general conceptual framework are then followed by a brief discussion of the promise of nonparametric techniques for the analysis of structural change issues and conclusions.

IMPLICATIONS OF THE STRUCTURAL CHANGE LINKAGE

The issue of structural change in consumption patterns has gained increasing attention from a variety of applied demand analysts, particularly agricultural economists. Significant trends in the consumption of red and white meat, whole versus low-fat fluid dairy products, fresh fruits and vegetables, fruit juices and soft drinks, food away from home, products with more convenience value added, and so on have challenged applied demand analysts for explanations and predictions. These challenges have raised questions concerning the adequacy of traditional research rules of thumb in the identification and estimation of the demand functions upon which much applied policy analysis is based.

Treating advertising as factors that shift marginal rates of substitution in consumption ceteris paribus clearly places advertising/consumption analysis within the context of structural change in consumption. The parallel issues concerning identification of technological change in production data are conceptually and empirically problematical. The identification of structural change in consumption confronts additional difficulties due to the fact that utility is indirectly revealed through consumption behavior in contrast to those that are directly observed as output, costs, or profits on the production side. This situation has led several notable agricultural economists to conclude that structural change in consumption cannot be identified given the available data and methodology (Haidacher 1983; Chavas 1983; Smallwood et al. 1986).

For the present purposes it is convenient to define structural change as a shift in the underlying and unobserved preferences that are assumed to be revealed in the demand functions summarizing the behavioral implications of these preferences given the current constraints (i.e., income and prices, household technology). If, for example, advertising successfully shifts the marginal rates of substitution between commodities ceteris paribus (in particular, independently of any changes in relative prices), then this would be considered as evidence of biased structural change. This could also imply a ceteris paribus change in the slope (and/or elasticities) of the demand curve as well as in the cross-price and income slopes (and/or elasticities), depending upon the specification.

There are several conceptual and empirical difficulties with this definition. Remember that indifference curves and the associated utility levels are not observed. This is much different from the parallel discussion of technical change in production isoquants where production (or profits or costs) are observed. Diamond et al. (1978) demonstrate the nonidentifiability of the elasticities of substitution and the bias in technical change in the absence of a priori knowledge of the structure of technical change. This identification problem means that more than one combination of technology and technical change hypotheses can rationalize a production data set. In most parametric specification/estimation this a priori structure becomes part of the explicit or (more commonly) implicit maintained hypotheses. This means that most parametric analysis of technology and technical change in production data is likely model dependent. When utility is not directly observed, in contrast to the production side, the difficulties are even worse.

A variety of reasonable hypotheses motivate structural change in consumption in addition to advertising: health and diet concerns, life-style changes, convenience dimensions of some products that are lacking in others, or quite generally, any factor that one might consider in a household production function context. Unfortunately, good proxies of several reasonably hypothesized sources of change in consumer preferences (e.g., diet and cholesterol concerns, or impacts of the federal dietary guidelines) are lacking. These data problems and the associated issues of distinguishing changes in household technology from changes in taste are quite common to the household production literature (Stigler and Becker 1977; Deaton and Muellbauer 1980, chap. 10).

In contrast to these general difficulties with the household production framework, better (though by no means ideal, particularly concerning proprietary brands) data on advertising and a considerable literature on the specification of advertising decay structures do exist (e.g., Bass and Clark 1972; Clark 1976; Jastram 1976; Kinnucan 1983 and 1985; and Simon and Arndt 1981). These factors increase the empirical feasibility of specifying a priori structure on the advertising sales response as means to identify structural change in consumption. While most applied analysis generally proceeds in this manner, it is very important how one proceeds. In particular, it is important how one explicitly (or as is quite frequent, implicitly) conceptualizes and specifies this structural change.

MODELING ADVERTISING AND PREFERENCES: A GENERAL CONCEPTUAL FRAMEWORK

The major research issue in modeling the impacts of generic and brand advertising for agricultural products concerns how to conceptualize and specify advertising and consumer preferences. If advertising changes preferences directly, then this can be modeled as a shift in marginal rates of substitution between commodities induced by some function of advertising. The household production model provides a very general conceptual framework for incorporating these effects.

For example, the general maximization hypothesis might be that consumers behave as if they

$$\begin{aligned} &\underset{X \geq 0}{\text{MAX}}\ U(Z_1, \ldots, Z_Q) \\ &\text{subject to: } Z_i = Z_i(X_i, B),\ i = 1, \ldots, Q \\ &\qquad\qquad\quad Y = \Sigma_i\, P_i X_i \end{aligned} \tag{7.1}$$

where the P_i are the prices of market goods, Y is a measure of total expenditures or income, and where components of the household production functions (Z_i) include a vector of purchased market goods (X_i), and B represents a vector of preference shifters that could include generic and brand advertising and/or other household technology shifters such as demographic and/or psychographic profiles, access to microwave ovens, and so on. The components of this B vector could also be interpreted as quality indicators of the various goods that in turn could be enhanced or diminished by advertising. Note that we generally have limited information on these variables and how they influence preferences, which is exactly the information required, explicitly or implicitly, to resolve the problems of identification discussed previously.

In the context of (7.1) the preference shift factors are hypothesized to influence the utility of market goods through the household production functions. Alternatively, the Z_i functions can be interpreted as "effective" quantity levels, which are functions of the observed quantities X_i "augmented" in a positive or negative fashion by the elements of the B vector. Under this interpretation the generalized augmentation hypothesis can be viewed as a special case of the household production model.

Substituting the household production functions (or, equivalently, the effective quantity specifications motivated by the augmentation hypothesis) into the utility function in (7.1) yields

$$U\{Z_1(X_1,B), \ldots, Z_Q(X_Q,B)\} \tag{7.2a}$$

or, in fully reduced form,

$$U(X_1, \ldots, X_Q; B) \tag{7.2b}$$

where the B in (7.2b) are interpreted as exogenous demand (preference) shifters as proxies for technology and/or taste changes. (See Stigler and Becker [1977] and Deaton and Muellbauer [1980, chap. 10] for these arguments.)

Assuming usual regularity conditions (i.e., well-behaved preferences, an interior solution), the reduced form demand functions associated with the maximization of (7.2b) subject to the standard budget constraint can be derived as

$$X_i = f(P_1, \ldots, P_Q, Y, B), \quad i = 1, \ldots, Q. \tag{7.3}$$

Note that this type of specification is frequently utilized to motivate the inclusion of demographic and other shift variables in commodity demand functions.

Phlips (1974, 181–83) employs fundamental demand matrix results to demonstrate that demand theory has additional implications for the demand functions in (7.3). This result also illustrates the basis of the structural change in consumption identification problems. In particular, demand theory implies that

$$S_B = -(1/\mu)*S_p^**V$$
$$= \{ -(1/\mu)*(\Sigma_q S_{iq}^* * V_{qr})\} \tag{7.4}$$

where S_B is the $(Q \times r)$ matrix of demand response (slopes) with elements $S_{ir} = \partial X_i/\partial B_r$ induced by the taste change factors $B = (B_r)$, μ is the marginal utility of income, S_p^* is the $(Q \times Q)$ Slutsky matrix of compensated price effects (slopes) with elements S_{iq}, and V is the $(Q \times r)$ matrix of second order cross-partial derivatives of the utility function with elements $V_{qr} = \partial^2U/(\partial X_q \partial B_r)$. While this result has been in the literature for more than 30 years (e.g., see Basmann 1956), Phlips's fundamental matrix presentation of this seminal insight is quite lucid and accessible.

Since the classical demand analytics basically generate the existence and properties of the marginal utility of income (μ) and the Slutsky matrix of compensated price effects (S_p^*), (7.4) is a very powerful result. On the one hand, (7.4) indicates that there are restrictions implied by demand theory on the effects of preference shift factors. While these restrictions are usually neither developed nor utilized in most of the applied literature, they can be tested for and, if found true, aid in the estimation of demand systems incorporating these shift factors.

On the other hand, (7.4) also indicates that the demand analytics of these shift factors require the additional knowledge of how these shift factors influence the curvature of the preference function [i.e., via the V_{qr} elements, the second-order cross-partial derivatives of $U(X)$]. That is, knowledge or assumptions about these second-order curvature properties imposes the structure necessary to resolve the identification problem. This additional information is either *implicitly* or *explicitly* imposed in the specification and estimation of commodity demand functions incorporating preference shift factors. Since this a priori structure is often latent, the relevant methodological questions concern how intelligently we proceed to specify, test, and explore the structure we must necessarily impose to make the measurement and estimation problem tractable.

Hanemann (1981) shows how the very general model in (7.1) and the results in (7.4) can be used to characterize a wide variety of demand situations. More important, the Hanemann results indicate that most of the recent theoretical and applied research in agricultural commodity advertising can be viewed as special cases of this general household production framework. Several conceptual insights emerge from this linkage and from the observation that we are basically attempting to model structural change effects similar to those encountered in production theory. In particular, we have a considerable store of known conceptual/empirical analytics and common insights to draw upon.

FOUR SPECIFICATION OPTIONS

Specification of the above general model involves choosing algebraic representations for consumers' preferences (direct or indirect utility or expenditure functions) and household production functions either in structural (equation [7.1]) or reduced form (equation [7.2b]). Equivalently, one can choose the algebraic representation of the reduced form demand functions as in equation (7.3). Note that this latter approach can also be characterized as choosing an algebraic representation for consumers' preferences since most popular demand function specifications such as the Rotterdam model have counterparts in explicit algebraic utility functions (see Barnett 1979, 1981; and Mountain 1988). Both approaches impose a priori structure on preferences and yield implicit or explicit restrictions of the form in equation (7.4) as a result of the identification problem.

Hanemann (1981) shows that there are at least three ways to generate empirical representations of this general conceptual model starting with an explicit formula for the direct (e.g. $U[X]$) or indirect (e.g. $V[P,Y]$) utility function (or through duality relationships for the expenditure function as well). His approach follows the convention of using a flexible functional form with standard neoclassical utility theory to model consumer preferences. Another major alternative not addressed by Hanemann but more prevalent in the applied agricultural commodity advertising literature is to directly specify demand functions incorporating advertising effects such as in (7.3) that are associated with (7.2a) or (7.2b). While theoretical consistency is still possible by imposing the restrictions (i.e., homogeneity, adding-up, and Slutsky symmetry) directly on these advertising augmented demand functions (e.g., as in the Rotterdam models), most of the agricultural advertising/consumption literature has followed a single equation approach. Hence, cross-equation restrictions such as Slutsky symmetry are seldom tested or imposed in these models. These models will be discussed more specifically in the fourth conceptual alternative.

SPECIFICATION OPTION 1

The first option is to add some terms involving components of B to the flexible functional form and then derive the corresponding formula for the ordinary demand functions (or compensated demand functions for an expenditure function). For example, letting $B = \{B_r\}$ and using the indirect translog utility function proposed by Christensen et al. (1975) would yield a formulation such as

$$\ln V(P,B,Y) = \Sigma_j\, \alpha_i \ln(P_j/Y) + 0.5\, \Sigma_i\Sigma_j\, \beta_{ij} \ln(P_i/Y)\ln(P_j/Y) + \Sigma_j\Sigma_r\, \pi_{jr} \ln(P_j/Y)\ln B_r. \tag{7.5}$$

In this case, the ordinary demand curves can be derived via Roy's identity, and the resultant budget share equations can be estimated by nonlinear least squares or maximum likelihood techniques. The work of Ellen Goddard in Chapter 13 of this book is an example of this translog indirect utility function approach. Amuah (1985) is another recent example of this approach from the agricultural commodity advertising literature using a generalized Box-Cox indirect utility specification.

SPECIFICATION OPTION 2

The second option concerns a more explicit formulation of the household production functions (Z_i) (or, equivalently, the generalized augmentation hypothesis) in (7.1). Pollak and Wales (1981) summarize two popular special cases of this approach as demographic translating and demographic scaling. More generally, as indicated by (7.1), these special cases of the augmentation hypothesis can be used to analyze the impacts of hypothesized taste change factors.

In the case of translating, the household production functions (or the effective quantity augmentations) are specified as

$$Z_i = X_i - \alpha_i(B) \tag{7.6}$$

where $\alpha_i(B)$ represents some function of the preference shift variables and X_i are the observed quantities (or the vector of market goods). Pollak and Wales (1981) show that the direct and indirect utility functions associated with the substitution of (7.6) in (7.1) take the form $U\{[X_1 - \alpha_1(B)], \ldots, [X_Q - \alpha_Q(B)]\}$ and $V\{P_1, \ldots, P_Q, [Y - \Sigma_j P_j \alpha_j(B)]\}$ respectively. The associated commodity demand functions can be derived in reduced form as

$$h_i(P_1, \ldots, P_Q, Y, B) = h_{\bar{i}}\{P_1, \ldots, P_Q, Y - \Sigma_j\, P_j\alpha_j(B)\} + \alpha_i(B) \tag{7.7}$$

where $h_i(\cdot)$ is the classical Marshallian demand specification of observed quantities and $h_{\bar{i}}(\cdot)$ is the Marshallian demand for "effective" or additively augmented (translated) quantities. Note that this additive form of the augmentation hypothesis implies that the demand impacts of the preference shift parameters basically act like income effects and generate linear specifications that are empirically convenient, hence popular.

In the case of scaling, the household production functions (or effective quantity augmentations) are specified as

$$Z_i = X_i / \alpha_i(B). \tag{7.8}$$

Pollak and Wales (1981) show that the direct and indirect utility functions associated with the substitution of (7.8) in (7.1) take the form $U\{X_1/\alpha_1(B), \ldots, X_Q/\alpha_Q(B)\}$ and $V\{P_1 * \alpha_1(B), \ldots, P_Q * \alpha_Q(B)\}$, respectively. The associated commodity demand functions can be derived in reduced form as

$$h_i(P_i, \ldots, P_Q, Y, B) = h_i^-(P_1 * \alpha_1(B), \ldots, P_Q * \alpha_Q(B), Y) * \alpha_i(B) \tag{7.9}$$

where $h_i(\cdot)$ is the classical Marshallian demand specification of observed quantities and $h_i^-(\cdot)$ is the Marshallian demand for effective or multiplicatively augmented (scaled) quantities. Note that this multiplicative form of the augmentation hypothesis implies that the demand impacts of preference shift parameters basically act like price effects and generate multiplicative specifications that are not as empirically convenient as the translating hypothesis. This specification is none the less quite common, particularly in the production/technical change literature.

Both of these formulations are completely general to both consumption and production analysis and to most hypothesized factors inducing structural (technical) change such as generic or brand advertising. Note that these specifications of the $\alpha(B)$ functions explicitly augment the utility (or output, profits, or costs, depending on the formulation) in a particular way, hence allow shifts in marginal rates of substitution between commodities (or inputs and/or outputs) to be induced by the hypothesized factors. Given an explicit representation of the direct or indirect utility (e.g., a translog specification) both of these augmentation hypotheses in the context of (7.1) generate the V_{qr} information implied by (7.4). Hence, the analytics on S_B can be derived using the traditional results on μ (marginal utility of income) and S_P^* (Slutsky matrix of compensated price effects).

The multiplicative augmentation (scaling) hypothesis is the most commonly used specification in the production literature. For example, a production function of the form $Q_t = A_t \cdot f(X_t)$ where Q_t is output, A_t is a technology index, and X_t is a vector of inputs, is a multiplicative output augmentation hypothesis commonly used to specify Hicks's neutral technical change (i.e., technical change that does not affect the marginal rates of substitution between inputs). In contrast, a specification such as $Q_t/A_t = f(B_t X_t)$ where B_t are technology indexes associated with inputs, allows for biased technical change under a multiplicative output *and* input augmentation hypothesis (that is, changes in the marginal rates of substitution between inputs and/or outputs). Such a bias in the structure of preferences with respect to commodity consumption seems exactly what much advertising hopes to achieve.

Examples of the scaling approach (either explicitly or implicitly) are numerous in the advertising/sales response literature. The "pure repackaging" model of Fisher and Shell (1968) explicitly formulates this type of advertising augmentation as disembodied taste change; that is, a change in tastes independent of any change in commodity quality but due to an information gain concerning the product [also see Deaton and Muellbauer (1980, chap.10)]. Further extensions and refinements of this model to advertising/sales analysis include Schmalensee (1972), Dixit and Norman (1978), and Nichols (1985). An insightful summary and overview of the Nichols model is provided in Ward and Chang (1985).

The additive augmentation (translating) hypothesis is found most frequently (seldom) in the consumption (production) literature. It is not clear why it is not more widely used, particularly since it is often easy to implement. Green (1985), for example, uses the Pollak and Wales (1981) demographic translating arguments to motivate an extension of the dynamic AIDS for the advertising/sales analysis along the lines of the next option.

SPECIFICATION OPTION 3

A third option for making this general household production framework operational is to combine options 1 and 2 by replacing the coefficients of the objective function (i.e., utility, indirect utility, or expenditure function) with a specified function of hypothesized structural change factors. The work of Green (1985) can be considered an example of this approach, where some parameters of the AIDS derived budget share equations

$$w_i = \alpha_i + \Sigma_j \mu_{ij} \ln P_j + \beta_i \ln(Y/P_0) \tag{7.10}$$

(with $\ln P_0 = \alpha_0 + \Sigma_k \alpha_k \ln P_k + 0.5 \Sigma_j \Sigma_k \mu_{kj} \ln P_k \ln P_j$) are allowed to vary with dynamic advertising effects. In particular,

$$\alpha_i = \alpha_i^* + \alpha w_{it-1}^a \tag{7.11}$$

where α is assumed constant across commodities and w_{it-1}^a is the lagged advertising budget share of the ith commodity. The impacts of advertising on own-, cross-price, and income elasticities that result from this theoretically consistent, flexible functional demand subsystem are easily derived. [See Green (1985, 83–86) for details.]

A Fourth Specification Alternative

While the previous three alternatives involve explicit specification of a preference (i.e., utility, indirect utility, or expenditure) function from which the implied demand functions are derived, it is also feasible to directly specify a demand function with advertising and other relevant ceteris paribus factors in reduced form as in equation (7.3) above. This conceptual approach characterizes most single equation applications found in the agricultural economics literature. Note, however, that this approach can easily accommodate the demand systems viewpoint by testing and/or imposing the restrictions implied by demand theory on the estimated demand functions. Examples of this approach would include development of constant elasticity of demand (CED) systems such as the double log, or the use of more flexible Rotterdam-type specifications.

This approach can also easily accommodate various augmentation hypotheses such as scaling or translating. As an example of incorporating the generalized augmentation hypothesis in this type of specification, consider the following CED system. The total differential of the directly specified demand function in (7.3) is

$$dX_i = \Sigma_q S_{iq}*dP_q + Y_i*dY + \Sigma_r B_{ir}*dB_r \tag{7.12}$$

where the S_{iq} are the Marshallian price slopes (i.e., $\partial X_i/\partial P_q$), Y_i is the slope of the associated Engel curve (i.e., $\partial X_i/\partial Y$), and the B_{ir} are the slopes of the preference shifters (e.g., generic and brand advertising and other shifters such as demographic characteristics).

Noting that the Marshallian slopes can be decomposed via the Slutsky identity, $S_{iq} = S_{iq} + X_q{*}Y_i$ where S_{iq} is the compensated price slope, and substituting in (7.12) yields

$$dX_i = \Sigma_q S^*_{iq}*dP_q + Y_i*(dY - \Sigma_q X_q*dP_q) + \Sigma_r B_{ir}*dB_r. \tag{7.13}$$

Note that this substitution yields compensated price slopes and real income effects (i.e., income is deflated by $\Sigma_q X_q{*}dP_q$). Taking logarithms as approximations to the differential terms in (7.13) yields a double-log CED system incorporating preference shifters.

$$\log X_i = \Sigma_q E_{iq}^* * \log P_q + E_{iy} * (\log Y - \Sigma_q w_q * \log P_q) + \Sigma_r E_{ir} * \log B_r \quad (7.14)$$

where E_{iq}^* are the compensated price elasticities, E_{iy} is the income elasticity, w_q are the budget shares of commodity q, and the E_{ir} are the demand impacts of the preference shifters in elasticity form. The demand restrictions in elasticity form are particularly easy to work with in this system given the compensated price slopes, E_{iq}^*. In particular, these restrictions are: $\Sigma_i w_i{}^*E_{iy} = 1$ (adding up); $\Sigma_q E_{iq}^* = 0$ (homogeneity); and $w_i{}^*E_{iq}^* = w_q{}^*E_{qi}$ (Slutsky symmetry). However, these restrictions can only be tested and/or imposed locally given that the budget shares, w_i, show up in the restrictions.

Taking first differences in logs as a discrete approximation to the differentials of the demand system in (7.13) and multiplying the result by w_i yields an alternative specification known as the Rotterdam model. In contrast to the double-log demand system, the restrictions can be tested and/or imposed globally in a Rotterdam context. Since the Rotterdam approach uses discrete approximations (i.e., first differences of logs) to the differentials in (7.13), it can be viewed as a local approximation to the underlying demand structure despite the fact that the demand restrictions can be tested and/or imposed globally. This is common to most popular flexible functional forms based on Taylor series approximations such as the AIDS and translog since they are local approximations around the point of the Taylor series expansion (see Barnett 1979, 1981; Mountain 1988).

The CED and Rotterdam models have been criticized for their lack of flexibility in modeling demand relations. However, these functional forms are flexible in the sense of not imposing a priori restrictions on the local Allen elasticities of substitution (AES). Their lack of flexibility arises from the fact that they are only locally consistent with the theory, or somewhat equivalently, that they become inflexible if they are forced to be globally consistent. Work by Wohlgenant (1982) suggests, moreover, that Rotterdam-type models are fairly robust in their ability to flexibly model structural change in meat demand relative to AIDS and translog demand systems. These demand systems have been shown to provide reasonable and robust approximations (particularly the Rotterdam) and are easy to implement empirically.

Incorporation of translating or scaling augmentation hypotheses in these demand systems is relatively straightforward. For scaling, note that (7.9) implies that the observed quantity demands, $h_i(P_1, \ldots, P_Q, Y, B)$, are augmented (scaled) by the $\alpha_i(B)$ functions to yield the effective quantity demands, $h_i(P_1{}^*\alpha_1(B), \ldots, P_Q{}^*\alpha_Q(B), Y)$. For simplicity, assume that $B = (g_1, \ldots, g_Q)$ and $\alpha_i(B) = \alpha_i(g_i)$ where g_i represents the stock of generic advertising for the ith commodity. This specification imposes a separable structure on the advertising stock effects. It does not imply that cross-commodity advertising effects are zero, but rather that they take a particular form similar to the impacts of separability on cross-price effects. The g_i stock of generic advertising measures would most likely be specified with a dynamic lag structure.

Taking the partial derivative of (7.9) with respect to the generic advertising stock of the rth commodity and converting to elasticity format yields

$$E_{ir} = \delta_{ir} * S_i + (E_{ir}^* + w_r * E_{iy}) * S_r \tag{7.15}$$

where δ_{ir} is the Kronecker delta, E_{ir} is the demand elasticity of the observed quantity of the ith commodity to generic advertising stock of the rth commodity [i.e., as defined in (7.14)], the S_j terms are the scale elasticities $\partial(\ln \alpha_j)/\partial(\ln g_j)$, and E_{ir}^*, w_r, and E_{iy} are defined in (7.14). Note that (7.15) restates (7.4) in elasticity format for the scaled CED system; that is, (7.15) states the empirical restrictions jointly implied by demand theory and the multiplicative augmentation (scaling) hypothesis in the context of the CED system stated in (7.14). These restrictions are empirically testable and, noting that the derivative of the budget constraint with respect to B_r implies that $\Sigma_i w_i * E_{ir} = 0$ if adding-up is to be satisfied, generate Q versus $Q^2 - Q$ advertising response parameters to estimate in the demand system.

This fourth specification option is quite general. Restrictions similar to (7.14), which incorporate translating or translating *and* scaling, or which are appropriate to Rotterdam demand systems, are easily derivable. Aside from generating degrees of freedom and potentially enhancing the efficiency of the parameter estimates (i.e., *if* the restrictions are true), the restrictions allow explicit testing of alternative implications of the theory under various augmentation hypotheses. For example, while multiplicative augmentation (scaling) is conceptually attractive and widely used in the specification of technical change in the production literature, if may not be appropriate for the analysis of advertising impacts on commodity demand systems. Hypotheses tests of alternative augmentation specifications can increase our inductively based insight for modeling these types of structural change factors.

SOME CAVEATS ON THE GENERAL CONCEPTUAL FRAMEWORK

One of the strengths of this general conceptual framework is that a system of demand functions that are fully consistent with the theory and incorporate advertising effects is analytically derived. In addition, cross-commodity advertising effects as well as theoretically consistent own- and cross-price effects are derived. This is very relevant for neoclassical welfare analysis of the impacts of advertising from a consumer and societal viewpoint. A considerable literature addresses these welfare concerns from what can be viewed as special cases of the above general conceptual model (e.g., Dixit and Norman 1978; Nichols 1985).

As with most conceptual models, the strength of the analytical results and insights are a function of the a priori structure that one is willing and able to impose on the problem. In the first three specification alternatives this involves explicit algebraic representations of the objective (utility, indirect utility, expenditure) function, the household production functions, the B vector, as well as the $\alpha(B)$ functions. In the fourth specification alternative this involves explicit algebraic representations of the demand and augmentation functional forms. Assumptions concerning the functional structure (separability) between the Z_i in the flexible objective function and the X_i and the elements of the B vector within the Z_i functions are generally required to make this conceptual framework empirically tractable. In general, without strong assumptions the resulting model can be quite nasty empirically. Obviously, the same can be said for most of the conceptual frameworks that we use empirically.

Aside from forcing one to be a little more explicit and thoughtful about the a priori structure often implicitly forced on our empirical results, this general conceptual framework can yield empirically testable restrictions implied by demand theory concerning the effects of preference shifters via the insights presented in (7.4). These types of additional restrictions on structural (technical) change parameters are generally absent from the available advertising/sales response literature. Hence, at the cost of being more explicit about the structure imposed on a problem to make it empirically tractable, we may be able to gain inductive insights and more precision in our estimates by exploring these types of restrictions more fully.

THE PROMISE OF NONPARAMETRIC TECHNIQUES

A major dimension of the structural change in consumption debate concerns the source(s) of changes in preferences. Most parametric specifications used in applied work impose functional structures not necessarily implied by the theory. This is particularly true concerning the estimation of structural change parameters when utility is not directly observed (i.e., the Diamond et al. [1978] arguments above). Recent developments in nonparametric demand analysis (Diewert and Parkan 1985; Varian 1982, 1983, 1985) are starting to be applied by agricultural economists (Chalfant and Alston 1986a and 1986b; Chavas and Cox 1987a). These techniques are also finding application in the analysis of technical change in production. (See Chavas and Cox 1987b, 1988a, 1988b.) If advertising impacts on consumption are appropriately modeled as explicit sources of structural change as argued above, then nonparametric applications of "revealed preference" theory hold considerable promise for the analysis of several (all too frequently) maintained hypotheses in the analysis of structural change.

These nonparametric methods do not generally assume functional forms for utility or demand functions. They also do not generate traditional price and income elasticity or slope estimates. One can, however, nonparametrically test for specific functional structure hypotheses such as separability over time periods and/or commodities. Similarly, one can test specific augmentation hypotheses such as translating (equation [7.5]) or scaling (equation [7.7]) motivated by the general conceptual framework. On the production side, recent work by Chavas and Cox (1987b, 1988b) indicates that it is possible to nonparametrically estimate additive augmentations to outputs and inputs in the context of profit or cost functions using linear programming techniques. The estimates can be interpreted as the minimum rates of bias in factor productivity that are consistent with the data. These types of issues are just beginning to be explored in the nonparametric literature and are considerably more complex on the consumption side for reasons discussed above. However, it appears that these nonparametric techniques may provide a powerful heuristic complement to more traditional parametric methodology.

CONCLUSION

The methodological bias of this chapter is largely that advertising/consumption analysis is strongly nested within the context of structural change in consumption issues. In order to make this bias more explicit, a conceptual framework has been presented in the spirit of household production theory through the use of generalized augmentation hypotheses following Hanemann (1981). The virtues of this approach are (1) explicit linkages to the technical change in production literature as a relevant source of insight and tools for advertising/consumption analysis; (2) explicit alternatives for incorporating structural change factors generally, and brand and generic advertising effects in particular, into demand systems that are consistent with neoclassical consumption theory; and (3) the feasibility of deriving structural inferences (restrictions) on the preference shift factors in the household production functions.

If the major premise developed in this chapter is correct [i.e., that consumption and production manifestations of the structural (technical) change are essentially parallel issues], then several research issues immediately arise from the linkage of advertising/consumption analysis to the technical change in production literature. In particular, serious identification and specification issues concerning structural change will need to be addressed in addition to the traditional specification issues such as maintained hypotheses about functional structure that are consistent with the data (e.g., separability over commodities and time periods, functional form of the objective and/or demand functions, specification of the advertising/sales decay structure, or form of the augmentation hypothesis). Wider use of nonparametric techniques to evaluate various maintained hypotheses prior to parametric estimation and wider use of theoretical specifications including those motivated by a generalized household production framework seem appropriate. These techniques hold promise as powerful heuristic tools to complement more traditional parametric approaches in the estimation of structural change. In addition, the need for time-series/cross-sectional (panel) data and better measures of advertising and consumer attitudes toward food and nutrition (e.g., attitude and tracking surveys) to support these techniques is essential.

Last, more rigorous and systematic evaluation of the research rules of thumb that we inherit and maintain seems appropriate if we are to operationalize demand theory for more than motivating consumption as a function of prices and income. Aside from *explicitly* imposing necessary a priori structure on the objective (utility, indirect utility, expenditure) or demand functions, the benefits of deriving structural inferences on the advertising effects motivated from the general conceptual model are primarily to generate inductive insights concerning a complex measurement problem. The derivation of additional structure (implied restrictions) about advertising effects using a theoretically consistent demand system can reduce data needs, facilitate hypothesis testing, and, one hopes, increase the statistical precision and our inductive understanding of these effects. This dimension of advertising/consumption response methodology is largely unexplored in the current empirical literature.

REFERENCES

Amuah, Alexander Kojo. 1985. "Advertising Butter and Margarine in Canada." Master's thesis, University of Guelph.

Barnett, W. A. 1979. "Theoretical Foundations for the Rotterdam Model." *Review of Economic Studies* 46:109–130.

_____. 1981. *Consumer Demand and Labor Supply*. Amsterdam: North Holland.

Basmann, R. L. 1956. "A Theory of Demand with Variable Consumer Preferences." *Econometrica* 24:47–58.

Bass, F. M. and D. G. Clarke. 1972. "Testing Distributed Lag Models of Advertising Effect." *Journal of Marketing Research* 10:298–300.

Chalfant, James A. and Julian M. Alston. 1986a. "Accounting for Changes in Tastes." California Agricultural Experiment Station Giannini Foundation Working Paper No. 402.

_____. 1986b. "Testing for Structural Change in a System of Demand Equations for Meat in Australia." California Agricultural Experiment Station Giannini Foundation.

Chavas, Jean Paul. 1983. "Structural Change in the Demand for Meat." *American Journal of Agricultural Economics* 65:148–153.

Chavas, Jean Paul and Thomas L. Cox. 1987a. "A Non-Parametric Analysis of the Structure and Stability of Consumer Preferences." Department of Agricultural Economics, University of Wisconsin–Madison. Staff Paper No. 268.

_____. 1987b. "A Non-Parametric Analysis of Productivity: The Case of U.S. and Japanese Manufacturing." Department of Agricultural Economics, University of Wisconsin–Madison. Staff Paper No. 273.

_____. 1988a. "A Non-Parametric Analysis of Agricultural Technology." *American Journal of Agricultural Economics* 70:303–310.

_____. 1988b. "A Non-Parametric Analysis of Productivity: The Case of U.S. Agriculture." Department of Agricultural Economics, University of Wisconsin–Madison. Staff Paper No. 281.

Christensen, L. R., D. W. Jorgenson, and L. J. Lau. 1975. "Transcendental Logarithmic Utility Functions." *American Economic Review* 65:367–383.

Clark, D. 1976. "Econometric Measurement of The Duration of the Advertising Effect on Sales." *Journal of Marketing Research* 13:345–357.

Deaton, A. and J. Muellbauer. 1980. *Economics and Consumer Behavior*. Cambridge: Cambridge University Press.

Diamond, P., D. McFadden, and M. Rodriquez. 1978. "Measurement of the Elasticity of Factor Substitution and Bias of Technical Change." In *Production Economics: A Dual Approach to Theory and Application*, vol. 2. Edited by D. Fuss and D. McFadden. Amsterdam: North Holland.

Diewert, W. E. and C. Parkan. 1985. "Tests for the Consistency of Consumer Data." *Journal of Econometrics* 30:127–147.

Dixit, A. K. and V. Norman. 1978. "Advertising and Welfare." *Bell Journal of Economics* Spring: 1-17.

Farm Foundation. 1985. *Research on Effectiveness of Agricultural Commodity Promotion*. Proceedings from Seminar Sponsored by Farm Foundation and USDA (AMS, ERS, FAS). Arlington, Va. April 9–10.

Fisher, Franklin M. and Karl Shell. 1968. "Taste and Quality Change in the Pure Theory of the True Cost-of-Living Index." In *Value, Capital and Growth: Papers in the Honor of Sir John Hicks*. Edited by J. N. Wolfe. Chicago: Aldine.

Green, Richard. 1985. "Dynamic Utility Functions for Measuring Advertising Responses." In *Research on Effectiveness of Agricultural Commodity Promotion*, 80–88. Proceedings from Seminar Sponsored by Farm Foundation and USDA (AMS, ERS, FAS). Arlington, Va. April 9–10.

Haidacher, Richard C. 1983. "Assessing Structural Change in the Demand for Food Commmodities." *Southern Journal of Agricultural Economics* July:31–38.

Hanemann, W. Michael. 1981. "Quality and Demand Analysis." In *New Directions in Econometric Modeling and Forecasting in U.S. Agriculture*. Edited by Gordon C. Rausser. New York: Elsevier/North Holland.

Jastram, Roy W. 1976. "A Treatment of Distributed Lags in the Theory of Advertising Effect on Sales." *Journal of Marketing Research* 13:345–357.

Kinnucan, H. W. 1985. "Evaluating Advertising Effectiveness Using Time Series Data." In *Research on Effectiveness of Agricultural Commodity Promotion* 105–122. Proceedings from

Seminar Sponsored by Farm Foundation and USDA (AMS, ERS, FAS). Arlington, Va. April 9–10.

_____. 1983. "Media Advertising Effects on Milk Demand: The Case of the Buffalo, New York Market." Department of Agricultural Economics Research Report 83-13. Cornell University, Ithaca, N.Y., February.

Kotowicz, Y. and F. Mathewson. 1979. "Advertising, Consumer Information, and Product Quality." *Bell Journal of Economics* 10(2):566–588.

Mountain, Dean C. 1988. "The Rotterdam Model: An Approximation in Variable Space." *Econometrica* 56:477–484.

Nichols, Len M. 1985. "Advertising and Economic Welfare." *American Economic Review* 75:213–218.

Phlips, L. 1983. *Applied Consumption Analysis*. Amsterdam: North Holland.

Pollak, R. and T. Wales. 1981. "Demographic Variables in Demand Analysis." *Econometrica* 49:1533–1551.

Schmalensee. 1972. *The Economics of Advertising*. Amsterdam: North Holland.

Simon, J. L. and J. Arndt. 1981. "The Shape of the Advertising Response Function." *Journal of Advertising Research* 20:11–28.

Smallwood, David M., Richard C. Haidacher, and James R. Blaylock. 1986. "A Review of the Research Literature on Meat Demand." Paper presented at the Symposium on the Demand for Meat, Charleston, South Carolina, October 20–21.

Stigler, G. and G. Becker. 1977. "De Gustibus Non Est Disputandum." *American Economic Review* 67:76–90.

Varian, H. R. 1982. "The Non-Parametric Approach to Demand Analysis." *Econometrica* 67:945–973.

_____. 1983. "Non-Parametric Tests of Consumer Behavior." *Review of Economic Studies* 50:99–110.

_____. 1985. "Non-Parametric Tests of Consumer Behavior." *Journal of Econometrics* 30:445–458.

Ward, Ronald W. and Julio Chang. 1985. "Theoretical Issues Relating Generic and Brand Advertising on Agricultural Commodities." In *Research on Effectiveness of Agricultural Commodity Promotion*. 89–104. Proceedings from Seminar Sponsored by Farm Foundation and USDA (AMS, ERS, FAS). Arlington, Va., April 9–10.

Wohlgenant, Michael K. 1982. "Structural Shifts in Demand for Meats: Taste or Quality Change?" Paper presented at the AAEA Annual Meetings, Logan, Utah.

CHAPTER 8

Measurement Issues in Empirical Demand Systems

Ronald A. Schrimper

THE QUALITY of empirical demand analysis depends not only on theoretical knowledge, analytical skills, and artistic abilities of researchers but also on the quality of available data. As noted by Edwards (1984, p. 28), "an important principle of agricultural economics research is what we study and what we conclude depends a great deal on what data are available." Concern about data is also evident in Gardner's comment (1983, p. 887) that "lack of attention to the ammunition we use in our sophisticated analytical artillery is an important impediment to the progress of agricultural economics as a science; and the silliness of our penchant to drop any bullet-shaped item down the barrel and see what happens when we pull the trigger is becoming dangerous."

The critical importance of data considerations on demand analysis is emphasized in this chapter. Part of the theme is the importance of understanding the data being analyzed, thinking about the underlying mechanisms responsible for the numerical values in basic data sets, and scrutinizing data for errors. Griliches (1985) noted that while it is useful to warn users about various imperfections and pitfalls of available economic statistics we must realize that available data are the main window we have to observe economic behavior.

The first part of this chapter discusses selected conceptual issues about data characteristics for particular kinds of demand analyses. The second part of the chapter addresses three specific empirical issues, namely, errors in data, missing data, and the effects of complex sample survey designs.

CONCEPTUAL ISSUES

A simple structural framework is useful for organizing the discussion of some conceptual issues related to desirable characteristics of data for demand analysis. The basic elements of the framework suggest that useful data sets must include information about some combination of quantities and/or expenditures, corresponding prices, incomes, or other influential variables observed across time or space in accordance with the theoretical notion of q_i or $E_i = f(p,I,Z)$ where q_i is the quantity of ith commodity or service per unit of time, E_i denotes expenditures on i^{th} commodity or service per unit of time, p_i is a vector of prices of goods and services, I is total income or expenditures, and Z is a vector of other variables influencing demand or expenditure behavior.

DEPENDENT VARIABLES

Certain questions about demand or consumption behavior may be answered by information about just the two dependent variables in the above framework. For example, questions concerning the trend in aggregate or per capita consumption of fresh versus processed dairy products or how the consumption of particular kinds of cheese varies among countries can be answered with just quantity or expenditure information. Similarly, individual firms may be interested primarily in what happens to sales of their products. These are relevant demand or consumption questions, but do not require answers from very elaborate data systems or analytical models.

As one moves to explore why certain changes in consumption or expenditure patterns have occurred, a more elaborate data system and mode of analysis may be required. A data set that can answer one question may be inadequate to answer a slightly different issue about the same subject. It is impossible to have one data system that can answer all questions. Characteristics that make data ideal for use in one situation can be excess baggage in other applications. This is why it is desirable to have different data sets that include similar information.

The multiplicity of consumption and expenditure data creates problems through the use of different methods for measuring more or less the same concepts in various data sets. For example, food consumption surveys collect information on actual quantities of food used by households for a specified period of time, whereas expenditure surveys obtain information about purchases. Although there is a distinct difference between the physical processes of purchasing and consuming, it probably does not make much difference which type of information is used for estimating demand parameters provided it is for the same population and observation period.

Aggregation Issues. Several aggregation issues arise for the dependent variables of the above structure in investigating any demand or expenditure question of interest. One is how to define appropriate categories of consumption or expenditures. Hicks's composite goods theorem indicates that commodities can be aggregated if their prices remain in a fixed relationship, but this concept is not all that useful in structuring a data-reporting system (Deaton and Muellbauer 1980). The desirable level of commodity aggregation or disaggregation ultimately depends on the type of question economic analysts are seeking to answer. Consequently, there is no universal way of structuring all data systems to address this issue satisfactorily. For example, data on the value of all dairy products over time may be sufficient to analyze the effect of major shifts in relative prices of food product groups. On the other hand, these aggregates may be completely meaningless if one is interested in studying changes in consumption or demand for particular dairy products.

Another aggregation issue is whether quantities or expenditures for a given product category are measured for individual consumers, households, or as disappearance for the entire economy. Frequently, economywide time-series aggregates are used for isolating price responses. On the other hand, cross-sectional individual or household observations are relied on more to estimate the influence of income and other factors on demand. Although it may not be too difficult to stretch theoretical models for individual consumers to represent the decision making of a representative household, the theory of aggregation over households for a market is rudimentary. Very quickly one reaches a fundamental dilemma in that theory is primarily micro-oriented, but many of our data are market-oriented. In working with market data, frequently the only satisfactory response to this issue is to assume that aggregate per capita values relate "to a representative consumer whose behavior is supposed to reflect the average behavior of the population" (Phlips 1974, p. 90).

Another issue closely related to aggregation concerns is whether reasonable answers can be obtained by estimating one demand equation apart from other categories of expenditures. Single equation approaches have provided many useful insights about demand parameters for policy purposes, but concern about consistency of estimates across commodities provides persuasive reasons to consider estimation of an entire demand system if sufficient data exist (Barten 1977). Even though estimation of demand systems is no longer as formidable as it once was because of computer technology, there are still many reservations. One is how much improvement occurs when a system of demand equations is estimated with aggregate market data when the supply side of the market is ignored.

It is clear that the gain of a more detailed system of demand equations is not a free good. As larger models are restricted to be consistent with theoretical models, caution must be exercised on how much the results are influenced by the choice of model versus the data used for estimation. This theme was emphasized by King (1979) in his presidential address to the American Agricultural Economic Association. In recent years considerable attention has been given to flexible functional forms to escape limitations of linear models. Wohlgenant (1984) discusses some efforts in this area pertaining to estimation of demand elasticities for food.

Independent Variables

A number of difficult issues regarding variables on the right-hand side of the above structure also exist in empirical studies. These relate primarily to prices, income, or other variables used to specify demand relationships.

Prices. In using time-series data or other information reflecting price variation, there is always a question of how many prices should be included in a demand equation. Although theory suggests that prices of all goods are relevant, pragmatism dictates some compromises. Criteria for making the choices are not clear.

When using cross-sectional data it is customary to concentrate on nonprice variables in demand or expenditure equations. When price data are available from cross-sectional surveys, however, they usually are not constant and thus there is a question of whether to ignore the information or how to interpret it. One interpretation is that variation in price may result from inclusion of different quality products in the same category. Consequently, estimates of price responses from such data may be confounded by quality differences. An alternative interpretation is that the price variation indicates the existence of separate markets. A recent review of the literature on this topic and an illustration of how household survey data can be utilized to estimate price elasticities is provided by Cox and Wohlgenant (1986). A very pragmatic problem, associated with cross-sectional price information, is the heterogeneity of container sizes or other units in which products are sold.

Income. Before a data set is selected for analysis considerable time should be spent in becoming familiar with the definitions used in collecting and reporting the information in order to know how to interpret results correctly. For example, questions used to obtain income information frequently differ among surveys and, consequently, different interpretations may be required. Also, reported income may or may not include capital gains, transfers, in-kind food or housing subsidies, and so on. Different income measures may be one reason various estimates of income effects are obtained with alternative data sets. When similar relationships are observed among a variety of data collected under widely differing circumstances, confidence increases. Thus, an eclectic approach to interpreting research results may be the most useful way to express our knowledge base about demand, consumption, and other economic phenomena. This view is consistent with thinking about empirical econometrics as being primarily for the purpose of developing essays of economic persuasion.

Even with all the empirical work in demand analysis on the influence of income differences on consumption patterns since Engel's pioneering work, there are still many unanswered questions. For example, should weekly, annual, or a more permanent measure of income be used? Some argue that total expenditure is a good proxy for permanent income. Use of total expenditure guarantees the adding up property for demand systems estimation will be satisfied. Using total expenditure, however, means that savings must be considered as one of the expenditure categories or treated separately from other expenditure decisions.

Regardless of whether a measure of income or expenditures is used, how should taxes, social security, and pension contributions be treated? Are these items to be subtracted from total income or expenditures or should some of them be treated as separate expenditure categories? Although a case can be made that consumers have little choice about some of these expenditure categories, it is also true that individuals with identical total incomes do not pay the same amount of taxes or compulsory deductions for social security and pension payments. It is clear that there is no easy way to define discretionary income or expenditures. For example, utility bills, mortgage, and/or auto payments may be as exogenous as some forms of taxes, at least in the short run, to the decision process about food expenditures. In a complete system of demand equations, however, expenditures on utilities, housing, and automobiles are objects of choice just like any other expenditure category.

Another issue about the income variable is if and how it should be adjusted for household size and/or composition differences. In some cases per capita or per adult equivalent measures of income are used instead of total household income. Given differences in measurement of such an important variable, it is little wonder that empirical estimates of income elasticities for identical products differ among studies. One example is analysis of 1965 USDA nationwide food consumption survey data. George and King's (1971) study reported an income elasticity for total food expenditures of .309. Using the same set of data but a different model, Salathe and Buse (1979) estimated an income elasticity for total food expenditures of .226. Both estimates are within the range of conventional wisdom about the magnitude of income elasticities for total food, but they differ by more than 25 percent. This magnitude of uncertainty might be acceptable for some issues but illustrates a challenge for further refinement of empirical demand analysis.

George and King (1971) used a linear in logs relationship between weekly per capita expenditures and per capita income after adjusting for the effects of household size (measured in 21-meal-equivalent units). Salathe and Buse's (1979) estimate of income elasticity was obtained by regressing weekly household expenditures on household income and several other variables to adjust for differences in household size, composition, and other characteristics. Both of these specifications are consistent with theoretical models of demand. This illustrates that theory does not provide all the answers in guiding the actual specification of a statistical model.

Other Factors. An excellent overview of recent studies of techniques for adjusting household expenditure and consumption data for differences in household size and composition is given by Deaton (1986). These include scaling and translating concepts (Pollak and Wales 1981). There does not seem to be any indication that one technique clearly dominates another, but there certainly is a tendency in applied work to do more than just express variables on a per capita basis when detailed information about household composition exists.

Household food use data usually require an adjustment for the number of meals eaten away from home and nutrient requirements of individuals in the household when comparing nutritional content of diets. Some of the assumptions underlying the customary process of adjusting data according to a 21-meal-equivalent basis have been examined (Smallwood and Blaylock 1984). In particular, they considered whether all meals contribute equally to dietary intake, whether average nutrient content of each meal type varies depending on whether it is an at-home or away-from-home meal, and whether individuals obtain nutrients in proportion to their dietary needs. Their study identified sufficient evidence to question the first and third assumptions. The second assumption appeared to be a less serious problem. After developing an alternative adjustment process that relaxed the two questionable assumptions, revised estimates of average nutrient intake did not indicate much difference from what resulted from the usual 21-meal-equivalent adjustment process. Thus, the 21-meal-equivalent adjustment method may be a reasonable scaling procedure for such survey data.

For other independent variables represented in the above conceptual model, there is considerable uncertainty about which ones are important and how to measure them appropriately. Given the small proportion of observed cross-sectional variation in household demand and consumption behavior explained by income and other conventional variables, it is obvious that more independent variables and interaction effects ought to be included in empirical models. A further reason for adding more explanatory variables is the extent to which biased estimates of parameters may occur, if important omitted variables are correlated with income and other included variables. Household production models suggest some new ways of thinking about reasons for including additional variables other than just picking up taste and preference differences among households. In particular, there is reason to include wage rates or other measures of the opportunity cost of time to reflect differences in full income and real prices of alternative commodities. In general, more work is required to give us a better guide for appropriate specifications of individual independent variables in empirical demand analysis.

ERRORS IN DATA

Every empirical analyst using economic data encounters problems with errors in data in one way or another. Several years ago several types of errors in economic data were documented (Morgenstern 1963). Many of Morgenstern's comments are still relevant with regard to using data originally collected for purposes other than to estimate demand systems. Griliches (1974) asserts that economists generally ignored Morgenstern's comments at the time they were made because the criticisms were rather indiscriminate and largely destructive. Griliches further argues that economists traditionally have had a rather ambivalent attitude about the data with which they work being subject to a variety of errors. He attributes much of the problem to a separation between data producers and data users in economics. Griliches argues that this was not necessarily the result of economics being a nonexperimental science. He notes that in some other nonexperimental areas, like astronomy or archeology, observations are made or collected directly by professionals themselves or individuals they supervise. On the other hand, ultimate responsibility for collection of much economic data is delegated to statistical bureaus or specialized survey firms.

Economic data generally are reported by people who may have no direct stake in the correctness of the information reported. Professionally trained collectors may gather information, but they usually must rely on responses elicited from others, who in some cases actually may have incentives to conceal correct information.

When experimenters or analysts themselves are responsible for data collection, they usually are keenly aware of potential data problems that might arise and have incentives to improve the quality of the information-gathering process. Increased interest in large microdata sets has created renewed awareness of problems with errors in data, since it is no longer possible to assume that errors associated with microunits cancel out in aggregate market data. Assuming that individual errors will be unimportant if a large enough sample is used may also provide a false sense of security.

Unfortunately, advances in computer technology that enable large data sets to be manipulated in sophisticated ways to estimate systems of demand equations may remove researchers even farther from the basic data.

An example illustrating the sensitivity of parameter estimates to errors in observations is provided by Ferguson's (1983) study of a disequilibrium model of the U.S. fed beef sector (Ziemer and White 1982). Ferguson reports that examination of the quarterly real per capita income series for 1965–79 obtained from Ziemer suggested a couple of suspicious values. The original data implied that real per capita income increased between the fourth quarter of 1972 and the first quarter of 1973 at an annual rate of 25 percent, only to decline almost as sharply two quarters later. Deletion of the two observations with suspicious income information resulted in more than a doubling of the coefficient for income and sizeable changes in other coefficients in the demand equation. Replacing the original values by interpolated estimates of real per capita income for the two suspicious observations produced estimates very similar to those obtained when the two observations were deleted.

Another dimension of the problem with errors in data noted by Griliches (1974) arises from the complexity of the phenomena being studied and measured. Thus, it is not surprising that measurement errors occur in empirical efforts involving concepts like expectations, permanent income, human capital, and reservation wages that are introduced in theoretical models of individual behavior underlying formulations of demand equations.

The major remedies for resolving errors in variables consist of being extremely careful in working with data and formally introducing exogenous information about relevant magnitudes of possible error variances as part of a formal estimation process. Most econometric textbooks provide a discussion of the problem and possible estimation alternatives.

MISSING DATA PROBLEMS

Anytime individual or household cross-sectional survey data are used for demand analysis, it is likely that the original set of observations will be missing some dependent as well as independent variables of interest. There are two reasons for this problem. One is that the original survey may not have asked the right set of questions to elicit all the information desired for a particular analysis. Jacob (1984) notes that users of available information are prisoners to the decisions about conceptualization made by those who are responsible for collecting the basic data. Another reason for missing data is that complete information about all the questions asked may not be available for research purposes. In certain situations respondents may have failed to provide sufficient information. In other instances the information may have been collected but not made available to researchers because of confidentiality criteria.

DEPENDENT VARIABLES

If purchase or use data for a large number of specific product categories are collected for a relatively short reference period, a number of zeros are likely to be observed. The nonnegativity characteristic of purchase or use data creates a special kind of "missing data" problem. The presence of zeros causes a problem in estimating any specific functional form of demand functions involving logarithms of quantities or expenditures. If the analysis is restricted only to observations with positive quantities or expenditures, resulting estimates may not be representative of the true parameters for all observations. On the other hand, even if particular functional forms that accommodate zero observations are used, appropriate adjustments in the estimating procedures may be required to satisfy usual statistical assumptions. A large number of zero observations suggest a variable with a truncated distribution. The basic problem with a truncated dependent variable is that the distribution of the error term in a linear model may not have a zero mean. If the latter condition is not fulfilled, ordinary least squares estimators will be biased.

The first procedure for combining information about nonpurchasers with actual purchases to estimate demand parameters was developed by Tobin (1958). A framework for handling many types of sample selection problems including the original question Tobin addressed was proposed by Heckman (1979). Heckman's framework makes it obvious why deletion of nonpurchasers from a sample of data as an explicit sample selection process can produce biased coefficients. It is clear that the actual selection process leading to the observations ultimately used in any analysis is a process depending not only on analyst decisions but also on those of respondents.

Respondents may self-select themselves out of a particular sample by refusing to provide any information about dependent or independent variables of interest. It is difficult to know exactly how interpretations of final estimates should be qualified because of this type of selection process. Fortunately, formal ways of incorporating the effects of sample selection have been developed. These essentially involve modeling the stochastic process responsible for determining particular values of the dependent variable, assuming complete information is available for independent variables. If information exists for all independent variables but only some of the dependent variables, one has a censored sample. If zero or some other arbitrary constant is recorded for some of the dependent variable observations, the censored sample would have a truncated dependent variable (Judge et al. 1982).

Heckman's approach for considering these issues is a two-equation model in which the availability of data for one of the dependent variables depends on the other dependent variable satisfying a particular inequality. An example of an application of this type of model for analysis of food expenditures is provided by Buse (1986). His model assumes that nonzero expenditures for a particular food category depend on each household initially making a decision to be a user or purchaser. To obtain consistent estimates of expenditure parameters for such a model, a two-step estimation process was used. The process involved first estimating a relationship explaining the purchase decision. The results of this equation were used to estimate the probability of being a purchaser, based on each household's characteristics. The probability of being a purchaser was then introduced into the expenditure equation as an additional independent variable. This procedure explicitly adjusted the parameters for the nonzero expectation of the error term because of the sample section process.

Although some variations of Tobin and Heckman's techniques for handling limited dependent variables may be very practical in cases of single demand equations, the possibilities at the present time for introducing these techniques into systems of demand equations are limited. The problems involved in trying to implement these techniques in such situations are discussed by Deaton (1986).

An alternative way of viewing zero expenditures has been proposed by Deaton and Irish (1984). Instead of modeling the occurrence of zero observations as a sample selection problem, they assume that zero expenditures or quantities occur with a certain probability representing the likelihood of purchase during a particular reference period. The magnitudes of the probabilities are affected by the length of the reference period. The results of implementing an appropriate estimating scheme consistent with this approach were somewhat disappointing according to the senior author. This may explain why he asserts that "the problem of dealing appropriately with zero expenditures is currently one of the most pressing in applied demand analysis" (Deaton 1986, p. 1809).

Independent Variables

The problem of missing information for independent variables becomes increasingly important with demand models that include a multitude of variables. The addition of each variable to a model increases the likelihood that a household will have supplied incomplete data for one or more of the variables of interest. Consequently, the number of observations with complete information in any data set is inversely related to the number of variables in demand equations. In situations in which original questionnaire information is unavailable, the only alternatives researchers have are to try somehow to fill in the gaps or discard observations with missing data. With larger data sets, discarding some observations may seem innocuous especially if there is reason to believe the missing data are randomly distributed throughout the population. A recent examination of missing income information in census data suggests, however, the latter assumption may not always be valid (Lillard et al. 1986). If the original data set is not very large, missing values may create a serious handicap for estimation and some attempt to fill in the gaps may be necessary.

Three different approaches for handling missing data for one of two independent variables in a simple linear model have been evaluated by Donner (1982). One alternative is to replace each missing value by the sample mean of the variable calculated from the remaining observations. Another alternative is to use predicted values for missing data using an estimated relationship among independent variables based on the observations with complete information. A third approach is to estimate element by element the sample covariance matrix for all variables using the observations with information for each pair of variables and then using the matrix of covariances rather than sums of squares and cross products to estimate parameters.

Donner (1982) derived the bias and variance of ordinary least squares estimators of the coefficients of the two independent variables when each of the three alternatives was used to fill in the gaps. The results were compared to what occurs if observations with missing data are deleted from the sample. It is not possible to generalize Donner's results to more complicated situations encountered in demand analysis. Nevertheless, there seems to be some support for the proposition that discarding observations and using only a subset of observations with complete information may not seriously impair results in all situations. Statements about the usual efficiency of ordinary least squares and other more complicated estimators may need to be modified, however, if observations are eliminated because of missing data. Generally, the worst alternative Donner examined was using complete information for all possible pairs of variables to estimate the sample covariance matrix before estimating the coefficients.

SAMPLING CONSIDERATIONS

Economists and others interested in empirical consumption and demand issues frequently rely on data that have been collected for purposes other than conducting analytical demand studies. This certainly was the case with some of the earliest empirical work in consumption and demand analysis. For example, early attempts to identify a systematic relationship between price and quantity were based on national production or trade statistics

(Stigler 1954; Johnson et al. 1984). Also, Engel's analysis of 153 Belgian household budgets, considered to be among the earliest empirical consumption analyses, was based on data collected by someone else (Stigler 1954). Thus, economists for some time have been using data in their demand and consumption studies that were collected for other purposes. Primary data collected by economists have also been used for demand analysis. Use of primary data in demand research does not appear to be as frequent as it was at one time. Part of the change may be a response to the increasing cost of collecting specialized information, more governmental and university regulations on dealing with human subjects, and more specialization of labor by researchers. Also, concern about the representativeness of specialized surveys of subpopulations has increased interest in using national data sources whenever feasible.

Whenever researchers collect their own data there are opportunities to exercise particular controls on the impact certain factors may have on the nature of the observations through the design of the experiment or sampling process. Much empirical demand work relies on the use of statistical procedures to adjust for the influence of particular variables that influence the outcomes in the set of data already collected. Evaluating alternative sets of data that might be used in a given situation is similar to considering what constitutes ideal conditions for designing experiments. In considering these issues it may become clear what compromises are made in selecting one sample of data over others. Researchers often have considerable choice about what data to use in their analyses. The limiting factor in producing good research is usually not data limitations but rather limited creativity and imagination in asking the correct questions and "teasing" new interpretations from data that generally are widely available. It is equally important to understand that it may be impossible to answer all questions of interest if the necessary experiment has not yet been conducted.

Sample selection issues are also encountered whenever subsamples of a larger sample are isolated for analysis. For example, if expenditure or demand behavior is analyzed for a subset of households with women in the labor force, sample selection problems may arise because the factors causing women to participate in the labor market may be related to consumption decisions. Therefore, expenditures may be affected by variables influencing labor force decisions as part of the sample selection process. Similarly, if one is interested in behavior of participants in a governmentally subsidized food program, behavior may be conditioned by certain factors differentiating participants from nonparticipants.

Stochastic Processes

Analysts using published time-series or available cross-sectional data to estimate demand relationships often do not spend as much time thinking about sampling problems as they would have if they had been responsible for collection. In using time-series data, frequently the only sampling issue to be determined is how many observations to use once a model has been specified. This decision usually does not involve specifying a stochastic process for physically selecting alternative observations from a designated population. Similarly, dependent and independent variables obtained from cross-sectional data sets frequently are used for estimating regression parameters without formal consideration of the stochastic processes determining the inclusion of particular observational units from some designated population.

Frequently, the only occasion when sampling issues are contemplated in demand analysis is when consideration is given to the statistical inference process. In the case of a set of time-series observations consisting of the finite population for a specified time period for a particular economy, a hypothetical set of repeated samples is conceptualized for purposes of thinking about the sampling distribution of the estimators. This is accomplished via the stochastic properties of an unobservable error term attached to the structural specification of the conceptual model of demand relationships discussed earlier in the chapter. The unobserved stochastic elements are assumed to account for the variation in the dependent variable of interest not attributable to explanatory variables in the model. A given set of dependent and independent variables is thereby assumed to be linked together by the particular outcomes of the selection process for the unobserved stochastic elements conditional on the specification of a particular model. The specific assumptions about the stochastic properties of the error term provide the basis for attributing statistical properties to alternative estimators of structural parameters. Adjustments in estimating procedures may be appropriate to obtain estimators with desirable statistical properties if the unobservable error terms are assumed to be the outcome of some process other than a random sample from an infinite population with a zero mean and finite variance.

When a sample of cross-sectional data is used to make an inference about parameters of an underlying structural model, two stochastic processes are actually involved. One random process is identical to that involved with time-series data. This is the random process assumed to determine the particular error term associated with each element of the finite population of observational units from which the cross-sectional sample of data is selected. The second stochastic process reflects how particular observational units in the finite population are selected for inclusion in the sample. The latter aspect often involves some type of stratified or multistage cluster selection process because of cost or other considerations involved in collecting data. This usually means that observations are selected with unequal probabilities.

Econometric textbooks concentrate discussion on the stochastic nature of the error term of structural models. In general, there is little discussion in such books about the process determining how particular observational units are selected from finite populations. Most discussions of econometric procedures initially proceed assuming exogenous variables of economic variables are nonstochastic or fixed. Sometimes this is interpreted erroneously to imply that usual statistical inference procedures are questionable if independent variables are not under control of the analyst. It also raises a question of how dependent variables of interest can be considered random and independent variables fixed if information about both is obtained from the same sample survey. For this reason, some individuals argue that economists and other researchers deceive themselves when traditional regression models are applied to sample survey data.

Johnston (1972) notes that fortunately there are two responses that can be used to address this issue. One position is that if the structural model correctly represents the expectation of the dependent variable conditional on specified values of the independent variables and if the error terms satisfy classical assumptions, then probability statements about confidence intervals and so on are still valid in terms of conditional probabilities for given values of the independent variables regardless of how the sample of observations was selected. The other alternative is to specify a joint likelihood function for the dependent and independent variables. As long as the distribution of the independent variables does not involve any of the unknown parameters to be estimated, maximum likelihood estimators are identical to regression estimators, assuming the independent variables are fixed.

Use of Sample Weights. When sampling weights accompanying each observation in a cross-sectional data set indicate that something other than a simple random sample is present, a decision must be made whether to consider the unequal probabilities in the estimation process.

Despite the appeal of the above two justifications for using standard regression procedures when a given set of data is not a simple random sample, some individuals still feel that selection probabilities should be incorporated into any estimation process using sample survey data. An explicit way of incorporating specific selection probability information about each observation in sample surveys in the estimation of linear regression coefficients is presented by Klein (1956). The procedure is a form of weighted regression. The general formula for the latter estimators is the same as for generalized least squares with a matrix of inverse probabilities replacing the inverse of the covariance matrix. These estimators can be represented as $Bv = (X'WX)^{-1}X'WY$, where X and Y have the usual interpretation as the set of independent and dependent variables, respectively, in a linear model, and W is a diagonal matrix with the diagonal elements being the reciprocal of the probability of selection of each observation. In the case of a simple regression model this amounts to

$$b_p = \sum_{i=4}^{n} x_i y_i / P_i \Big/ \sum_{i=1}^{n} x_i^2 / P_i \tag{8.1}$$

where x_i and y_i are deviations from the respective weighted means ($\bar{X}_w$ and $\bar{Y}_w$) as specified by

$$X_w = \sum_{i=1}^{n} X_i / P_i \Big/ \sum_{i=1}^{n} 1/P_i \quad \text{and} \quad Y_w = \sum_{i=1}^{n} Y_i / P_i \Big/ \sum_{i=1}^{n} 1/P_i \tag{8.2}$$

and P_i represents the probability of selection of the ith observation. In the case of a simple random sample with replacement, P_i would be identical to n/N (where n is sample size and N is the size of the finite population) and the weighted formula would be the same as the unweighted formula.

Descriptive versus Analytical Inference. Another case in which P_i would be identical is if the sample consisted of the entire finite population. In this case the estimator would be

$$b_n = \sum_{i=1}^{N} x_i y_i \Big/ \sum_{i=1}^{N} x_i^2 \tag{8.3}$$

Equation 8.3 indicates the least squares estimator that would occur if the entire finite population had been available to calculate an estimate of the single unknown structural regression coefficient. Some analysts view the latter expression as the major object for statistical inference in the absence of being able to observe the entire finite population. Pfeffermann and Smith (1985) characterize this as descriptive inference. It consists of inference about a known *function* (not value) of the finite population. This is in contrast to analytic inference, which refers to inference about parameters of a model. Regression analysis is usually viewed as a tool for analytic inference, but some individuals have made a strong case for a descriptive approach to regression analysis when using sample survey data (Kish and Frankel 1974). The descriptive approach is concerned with defining appropriate combinations of sample data that would approach the value of the function that would occur if the entire finite population had been observed. A weighted (or more specifically probability weighted) estimator is the appropriate formula to estimate what the regression coefficients of some unknown structural parameters would be if one had observed the entire finite population. Exact values of the unknown parameters of the underlying structural model of interest would not occur, however, even if the entire finite population had been observed because of the sampling process associated with unobserved structural error terms.

Probability weighted formulas arise because sample survey specialists usually are concerned with procedures for using sample information to estimate particular statistics describing a finite population. This is analogous to the case of appropriate probability weighting sample data when estimating a mean or frequency distribution of a finite population. In the latter cases, however, the objective is to estimate particular fixed characteristics of a finite population rather than structural parameters of a theoretical relationship. Tobin (1958) provides the following argument for basing his analysis on simple counts of sampled spending units without allowing for the fact that the sampling design gave some spending units greater probability of being included in the sample. He said:

> The purpose of this example is not to estimate population frequency distributions, but only to examine the relationship of durable goods expenditure to age and liquid asset holdings within this sample. It is not necessary to consider here how the relationship exhibited in this sample differs from the one that would be exhibited in a complete enumeration. But it may well be that the sample gives unbiased estimates of the parameter of the relationship even though it gives biased estimates of the separate frequency distribution of the variables. (p. 31)

Thus, ordinary unweighted least squares procedures are appropriate to use with a sample of observations with unequal probabilities of selection to make analytical inferences about structural parameters provided the error terms in the underlying model are independent and have a zero mean and constant variance. As long as the sample selection process does not affect the mean of the error term, ordinary least squares estimators would still be unbiased. If the independence or constant variance assumption of the error terms appear questionable, some type of generalized least squares estimators would be more efficient.

If there is interest in an estimator that would be identical to what one would get if the entire finite population were available, then a probability weighted estimator is appropriate. Even though the formula for obtaining probability weighted estimators is a type of generalized least squares estimator, caution must be exercised in calculating the variance of the resulting estimator. Klein (1956) noted that the variance of the probability weighted estimator differs from that of weighted estimators that adjust for heteroskedasticity. Holt et al. (1980) also raise a caution flag about this issue, especially when probability weighted least squares estimates are obtained using standard computer packages for weighted regression. They note that correct values of the parameter estimates but erroneous estimates of the variance of probability weighted estimators are likely to occur.

An interesting use of the difference between the probability weighted and unweighted least squares estimators of a linear model has been proposed (DuMouchel and Duncan 1983). They argue that a difference in the two estimators can be used as a basis for searching for a more appropriate model specification. The basic idea is that differences between weighted and unweighted estimators reflect specification bias that would not exist if one had a correct specification of the underlying model.

CONCLUSION

Several issues regarding the availability and usability of data for econometric estimation of demand models have been highlighted in this chapter. Difficulties in empirically operationalizing theoretical models of consumer behavior arise for many reasons. The widespread tendency of economists to estimate demand relationships based on data collected for other purposes creates some problems but at the same time permits specialization of effort. To avoid erroneous inferences about demand behavior based on data collected for other purposes, extra effort must be spent to understand how various concepts were defined when the information was originally collected. Another source of difficulty for empirical demand analyses is that the current theoretical base of economics does not provide a detailed blueprint for specification of empirical demand relationships. Consequently, researchers still have to exercise considerable judgment in using any particular data set to explore a particular question of interest.

Some comments about data errors, missing data, and complex sample designs conclude the latter part of this chapter. Advances in computer technology enabling the analysis of larger data sets from complex sample surveys unfortunately mean it is difficult for researchers to become familiar with the basic data. Hence, extra caution about errors in data appears warranted. Missing information or a large number of zeros for dependent variables of interest can be rationalized by a number of alternative models. Statistical estimation of such models requires explicit ways of combining different types of information about the dependent variables of all observations.

Several alternatives for handling missing information for independent variables of interest in demand models are possible. There is limited support for the proposition that discarding observations with missing independent variables may not be too bad in many situations.

Complex sample designs for collecting survey data lead to observations with unequal probabilities of selection. When estimating regression coefficients of demand equations for purposes of analytical rather than descriptive inference, however, the unequal probabilities of selection of the sample observations can be ignored.

REFERENCES

Barten, Anton P. 1977. "The Systems of Consumer Demand Functions Approach: A Review." *Econometrica* 45:23–52.

Buse, Rueben. 1986. "Is the Structure of the Demand for Food Changing? Implications for Projection." In *Food Demand Analysis, Implications for Future Consumption*. ed. Oral Capps, Jr., and Benjamin Senauer. Blacksburg, Va.: Department of Agricultural Economics, Virginia Polytechnical Institute and State University.

Cox, Thomas L. and Michael K. Wohlgenant. 1986. "Prices and Quality Effects in Cross-Sectional Demand Analysis." *American Journal of Agricultural Economics* 68:908–919.

Deaton, Angus. 1986. "Demand Analysis." In *Handbook of Econometrics*. ed. Z. Griliches and M. Intriligator, 1767–1839. New York: North Holland Press.

Deaton, Angus and John Muellbauer. 1980. *Economics and Consumer Behavior*. Cambridge: Cambridge University Press.

Deaton, Augus and M. Irish. 1984. "A Statistical Model for Zero Expenditures in Household Budgets." *Journal of Public Economics* 23:59–80.

Donner, Allan. 1982. "The Relative Effectiveness of Procedures Commonly Used in Multiple Regression Analysis for Dealing with Missing Values." *The American Statistician* 36:378–380.

DuMouchel, William H. and Greg J. Duncan. 1983. "Using Sample Survey Weights in Multiple Regression Analyses of Stratified Samples." *Journal of the American Statistical Association* 78:535–543.

Edwards, Clark. 1984. "Wheat Price: Past and Future Levels and Volatility." *Agricultural Economics Research* 78:535–543.

Ferguson, Carol A. 1983. *An Evaluation of Disequilibrium Model*. Agricultural Economics Research 83-27. Ithaca: Department of Agricultural Economics, Cornell University.

Gardner, Bruce. 1983. "Fact and Fiction in the Public Data Budget Crunch." *American Journal of Agricultural Economics* 65:882–888.

George, P. S. and G. A. King. 1971. "Consumer Demand for Food Commodities in the United States with Projections for 1980." Giannini Foundation Monograph No. 26. California Agricultural Experiment Station.

Griliches, Zvi. 1974. "Errors in Variables and Other Unobservables." *Econometrica* 42:971–998.

_____. 1985. "Data and Econometricians—The Uneasy Alliance." *American Economic Review* 75:196–200.

Heckman, James J. 1979. "Sample Selection Bias as a Specification Error." *Econometrica* 47:153–161.

Holt, D., T. M. F. Smith, and P. D. Winter. 1980. "Regression Analysis of Data from Complex Suveys." *Journal of the Royal Statistical Society, Ser A.* 143:474–487.

Jacob, Herbert. 1984. *Using Published Data: Errors and Remedies*. Sage University Paper Series on Quantitative Applications in the Social Sciences, 07-042. Beverly Hills and London: Sage Publications.

Johnson, Stanley R., Zuhair A. Hassan, and Richard D. Green. 1984. *Demand Systems*

Estimation, Methods and Applications. Ames: Iowa State University Press.

Johnston, J. 1972. *Econometric Methods*. 2d ed. New York: McGraw-Hill.

Judge, G. G., R. C. Hill, W. E. Griffiths, H. Lutkepohl, and T. C. Lee. 1982. *Introduction to the Theory and Practice of Econometrics*. New York: Wiley.

King, Richard A. 1979. "Choices and Consequences." *American Journal of Agricultural Economics* 61:839–848.

Kish, L. and M. R. Frankel. 1974. "Inference from Complex Samples." *Journal of the Royal Statistical Society* B 36:1–37.

Klein, Lawrence R. 1956. *A Textbook of Econometrics*. Evanston: Row Peterson.

Lillard, Lee, James P. Smith, and Finis Welch. 1986. "What Do We Really Know About Wages? The Importance of Nonreporting and Census Imputation." *Journal of Political Economy* 94:489–506.

Morgenstern, Oskar. 1963. *On the Accuracy of Economic Observations*. 2d ed. Princeton: Princeton University Press.

Pfeffermann, D. and T. M. F. Smith. 1985. "Regression Models for Grouped Populations in Cross Section Surveys." *International Statistical Review* 53:37–60.

Phlips, Louis. 1974. *Applied Consumption Analysis*. Amsterdam: North-Holland.

Pollak, R. A. and T. J. Wales. 1981. "Demographic Variables in Demand Analysis." *Econometrica* 49:1533–1551.

Salathe, Larry E. and Rueben C. Buse. 1979. "Household Food Consumption Patterns in the United States." Technical Bulletin, No. 1587. Washington, D.C.: U.S. Department of Agriculture, Economics, Statistics, and Cooperative Services.

Smallwood, David and James Blaylock. 1984. "Scaling Household Nutrient Data." *Agricultural Economics Research* 36:12–22.

Stigler, George J. 1954. "The Early History of Empirical Studies of Consumer Behavior." *Journal of Political Economy* 62:95–113.

Tobin, James. 1958. "Estimation of Relationships for Limited Dependent Variables." *Econometrica* 26:24–36.

Wohlgenant, Michael K. 1984. "Conceptual and Functional Form Issues in Estimating Demand Elasticities for Food." *American Journal of Agricultural Economics* 66:211–215.

Ziemer, Rod F. and Fred C. White. 1982. "Disequilibrium Market Analysis: An Application to the U.S. Fed Beef Sector." *American Journal of Agricultural Economics* 64:56–62.

140-77

CHAPTER 9

Alternative Data Sources for Demand Estimation

D11 D12 C81 US D12

Rueben C. Buse

A WIDE VARIETY of data sets can be used in estimating and/or analyzing demand. They can be categorized in two dimensions: cross-sectional versus time-series and experimental versus nonexperimental. Nonexperimental cross sections and time series and variations on the two data types are the most frequently used data sets. The experimental approach systematically varies the independent variables of interest such as price or income, and controls the influence of other exogenous variables on the dependent variable. Experimental data have fewer technical problems because of the controls on variation and covariation of the independent variables. In practice, the experimental approach has been limited to very specific research problems because of the difficulty of setting up reliable controls in real world situations at reasonable cost. There are no publicly available experimental data sets useful for demand estimation.[1]

In contrast, no control is exercised over the variation in the independent or over their covariation in the nonexperimental approach. The nonexperimental approach uses whatever changes occur naturally in the environment from which the data are collected, recording the data ex post. The majority of nonexperimental data available are historical data series collected over time (time-series) or surveys (cross-sectional). The next section elaborates further on these two major data types.

Data series can also be classified as public or private according to the type of organization producing them and/or their general availability to researchers. Public data sets are usually generated by government agencies and are readily available to the researcher at a small cost. In the United States the most widely used and readily available public data used in demand analyses have been designed, developed, and maintained by the United States Department of Agriculture (USDA), the Department of Commerce through the Bureau of Labor Statistics (BLS), and special research studies by universities. Private data sets are generated as a source of income to private firms and are much more difficult to acquire, usually at substantial cost. Those data sets are usually for special purposes. A number of both public and private data sets useful for demand analysis will be described in this chapter.

In almost all cases, whether public or private, the data were not designed for the purpose of using them in demand analyses. Therefore, in addition to describing the data series available for demand analysis it is also important that a potential user know more about the characteristics of the different types of data. In this chapter the next section elaborates on the characteristics of nonexperimental cross-sectional and time-series data sets and the following section examines conceptual and statistical issues in the various data types. Subsequent sections present detailed discussions of the various data sources and systems that are currently useful in demand analysis. The final section discusses problems the user must be aware of in using the various data sets in demand analysis.

The discussion focuses on U.S. data sets suitable for demand analysis but there are many comments and insights throughout that will be useful to researchers in other areas of the world. Several sections describe problems, offer comments and suggestions, and provide insights and cautions useful to anyone interested in generating and/or using data for demand analysis. The Food and Agricultural Organization of the United Nations (FAO) has published a comprehensive listing of the household food consumption and expenditure surveys available for each country in the world (1981, 1986). Other FAO publications publish annual time-series data on food production and trade for each country of the world.

TYPES OF DATA

A structural framework is useful for organizing a discussion of the nonexperimental data types that can be used in demand analysis. A conceptual demand equation(s) is used. Assuming an individual maximizes utility, demand theory leads to the following demand equation for a particular commodity at time period t for individual i

$$Q_{it} = b_0 + b_1 Y_{it} + BP_{it} + CZ_{it} + e_{it} \tag{9.1}$$

where Q_{it} = quantity of the good consumed/purchased, Y_{it} = income for individual i, P_{it} = vector of prices, Z_{it} = other variables influencing demand behavior, e_{it} = stochastic error term, b_0, b_1 = parameters to be estimated, B = vector of price parameters to be estimated, and C = vector of parameters on demand adjusters.

Abstracting from the many operational and estimation problems, equation (9.1) illustrates the basic elements of the analytical models used to study demand. It is clear that to estimate (9.1) the researcher must include data on quantity, prices, income, and other variables influencing the level of Q.[2]

The ideal way to estimate equation (9.1) is the experimental approach, that is, vary one variable at a time and observe the impact on Q_{it}. The next best way is to observe a representative sample of consumers over time, recording changes in quantity and the corresponding levels of the relevant economic variables on the right-hand side of (9.1)—a quasi-experimental approach. The same units, observed repeatedly, serve as their own controls.

In practice both procedures have been impractical. Controlling the prices paid by individuals and their incomes is not possible for very long. Following individuals over time is extremely expensive and fraught with statistical problems. More important, for individuals the variables in (9.1) are difficult to measure since most consumption takes place in a household environment. The researcher must use the appropriate theoretical modifications to (9.1) and one of several variants on the ideal data set. The three most common are time-series data, cross-sectional data, or panel data. A fourth type of data, the result of recent technological advances in food store inventory control and checkout procedures, is scanner data.

TIME-SERIES DATA

Time-series data measure aggregates of the variables in (9.1) during successive time periods. The time period is often a year but it can be a quarter, month, or week. Sometimes longer time periods such as two, three, or five years are used. The changes in quantity are related to changes in the variables on the left-hand side of (9.1) over time.

For equation (9.1) a time-series sample of T observations for N individuals is

$$\begin{array}{lllll} \text{Period 1:} & Q_{.1}, & Y_{.1}, & P_{.1}, & Z_{.1} \\ \text{Period 2:} & Q_{.2}, & Y_{.2}, & P_{.2}, & Z_{.2} \\ & \cdot & \cdot & \cdot & \cdot \\ & \cdot & \cdot & \cdot & \cdot \\ & \cdot & \cdot & \cdot & \cdot \\ \text{Period } T\text{:} & Q_{.T}, & Y_{.T}, & P_{.T}, & Z_{.T} \end{array} \tag{9.2}$$

For each time period, the observations on Q and Y are aggregated across

N individuals, P_T is a vector of average observed prices relevant to the demand equation being estimated, and Z_T is a vector of intercept/slope shifters reflecting changes in the conditions affecting demand over time or variables describing changes in the distribution within the group being aggregated. The prices are average prices but may not be the actual prices paid by any of the consumers in the aggregate. The matrix Z may include a measure of the number of individuals reflected in the aggregate or the data may be normalized to a per capita basis. The parameters, b_0, b_1, B, and C are estimated from changes in the data across time.

In practice, the data on quantity and prices in (9.2) are from markets, that is, total quantity moving through the market in a specified period of time, the corresponding average prices prevailing in that period. The data on quantity and prices may be generated at various market levels including farm, assembly market, central market or from processors, wholesalers, or retailers. Public historical data series collected and published by public agencies are the most readily available sources of these data. An example is annual USDA disappearance data. The researcher must exercise care in matching the appropriate price and socioeconomic data with the quantity in the data series.

Another type of time-series data reports product movements within a marketing channel, usually at the processor or wholesale level. Most are collected by private agencies and are not available without cost to researchers outside the firm. An example is the Statistical Area Market Information (SAMI). SAMI reports product shipments from distribution warehouses to retail stores every four weeks.

The theory upon which equation (9.1) is based is for an individual consumer while the data are from the market. An assumption of the time-series approach is that, across the T time periods, the distribution of individuals in the market level aggregates and their behavior is constant. Thus, the use of time-series data to estimate the parameters of (9.1) imposes the assumptions that the individual theory holds for market level demands and that b_0, b_1, B, and C estimated from market level data are the same as the parameters of a single representative consuming unit. [For more details on the aggregation problem, see Intriligator (1978, 233ff).] Variables reflecting changes in population distributions are added to Z to control for such effects.

In most time-series data the assumption of behavior homogeneity across time is also violated, that is, the parameters b_0, b_1, B, and C are not constant over time. External events disrupt the continuity of the market structure. War years, depressions, oil crises may have changed how the consumer responds to the same independent variable. Finding variables that can reflect heterogeneity across time is difficult. Trends, changes in the population distribution, and binary variables are added to the Z vector in equation (9.1) to account for temporal heterogeneity.

Cross-Sectional Data

An alternative approach to obtaining parameters for (9.1) is to observe a set of consumers at one period of time and relate differences in the quantity consumed to differences in levels of the independent variables across the observed units. This method is called the "cross section" approach. Cross-sectional data measure a particular set of variables at a given time for different individuals.

A cross section of N observations (consuming units/households) at time period t is

$$\begin{array}{lllll} \text{Unit 1:} & Q_{1t}, & Y_{1t}, & P_{1t}, & Z_{1t} \\ \text{Unit 2:} & Q_{2t}, & Y_{2t}, & P_{2t}, & Z_{2t} \\ & \cdot & \cdot & \cdot & \cdot \\ & \cdot & \cdot & \cdot & \cdot \\ & \cdot & \cdot & \cdot & \cdot \\ \text{Unit } N\text{:} & Q_{Nt}, & Y_{Nt}, & P_{Nt}, & Z_{Nt} \end{array} \tag{9.3}$$

The data are usually generated by a survey on consumption or expenditures. Generally, the surveys record data on all units for the same time period and "t" is dropped. In general, cross-sectional surveys of food demand or consumption have not included price information. Consequently, researchers using data set (9.3) have assumed prices constant in estimating (9.1). Estimating the parameters of equation (9.1) with cross-sectional data also assumes that the jth individual with values for the independent variables of Y_{jt} and P_{jt} behaves like the kth individual with values of Y_{kt} and P_{kt} if the jth's values were to change to Y_{kt} and P_{kt}. Thus, cross-sectional data assume homogeneity across individuals, that is, b_0, b_1, B, and C are the same or can be adjusted to reflect differences across sample groups. The variables included in Z are used to adjust for *known* behavioral heterogeneities. Usually, the variables in Z are sociodemographic descriptors of the consuming unit such as age of the head, number of people in the unit, marital arrangement, wealth, expectations on income and prices. Household consumer expenditure surveys by the BLS and household food consumption surveys by USDA are examples of cross-sectional data useful for demand analysis.

Pooled Time Series and Cross Sections

Sometimes cross-sectional and time-series data are merged or "pooled" to take advantage of the strengths of both data types. The result is a time series of cross sections or a cross section of time series. In pooling time-series and cross-sectional data the information from both is used to estimate the demand model. For example, the income coefficient (b_1) would be estimated from cross-sectional data conditioned on family size, race, education, and geographic location while holding (assuming) prices constant. Given the estimated income coefficient(s), time-series data are used to estimate the demand parameters associated with prices. This method is sometimes used to remedy the multicollinearity problem among the independent variables in time-series data. (See the section on statistical problems for more details.)

PANEL DATA

The third type is panel data. Panel data are a special type of pooled cross-sectional and time-series data. They contain observations of the same consuming unit over several time periods. They are superior data for studying individual demand behavior because the individual observation is used as its own control.

Panel data would include a sample of N observations over T time periods.

$$
\begin{array}{lll}
\underline{\text{Period 1}} & & \\
& \text{Unit 1:} & Q_{11},\ Y_{11},\ P_{11},\ Z_{11} \\
& & \vdots \\
& \text{Unit } N\text{:} & Q_{N1},\ Y_{N1},\ P_{N1},\ Z_{N1} \\
\underline{\text{Period 2}} & & \\
& \text{Unit 1:} & Q_{12},\ Y_{12},\ P_{12},\ Z_{12} \\
& & \vdots \\
& \text{Unit } N\text{:} & Q_{N2},\ Y_{N2},\ P_{N2},\ Z_{N2} \\
& \vdots & \vdots \\
\underline{\text{Period } T} & & \\
& \text{Unit 1:} & Q_{1T},\ Y_{1T},\ P_{1T},\ Z_{1T} \\
& & \vdots \\
& \text{Unit } N\text{:} & Q_{NT},\ Y_{NT},\ P_{NT},\ Z_{NT}
\end{array}
\tag{9.4}
$$

Panel data useful to demand analysts are not widely available since they are expensive to obtain and usually designed for specialized research objectives. The major difficulty with panels is sample attrition, that is, with each succeeding time period a percentage of units drop out of the survey because they stop cooperating, move, or otherwise disappear. In practice, "pseudopanels" are often used. In the pseudopanel there is no attempt to maintain the individual units in the sample over time. In this approach a set of consuming units is observed and measured at one or more time periods and then replaced by comparable units. In either case, the problems of panel data are a combination of those of both time-series and cross-sectional data.

Using data set (9.4) the coefficients on the explanatory variables in (9.1) are subject to both time and individual effects. The researcher using panel data must include variables in Z that allow for the possibility of differences in behavior over cross-sectional units as well as differences in behavior over time for a given cross-sectional unit. The appropriate variables and estimation procedure depend upon the maintained hypothesis for the affected coefficients. One can assume the coefficients vary over individuals, over time, or both (binary variable models). Alternatively, one can assume model coefficients are random variables with components for individuals, time, or both (error-components model). [For details see Judge et al. (1980, chap.8).]

The oldest nationwide household panel is the MRCA-Panel developed by the Market Research Corporation of America. It is a private data set established in 1939. The most recent public panel data is the BLS Consumer Expenditure Survey begun in 1980. It is a pseudopanel in that households rotate out of the sample and are replaced periodically.

Store panels are another type of panel data. Continuous data from a sample of retail stores were developed because of the high cost of household panels. The A. C. Nielsen Company is a private company reporting data on sales of food chains obtained using store panels. In this approach store sales are audited every 60 days to obtain information on product sales during the past 30 days. The results measure retail sales by determining the quantity of product delivered to the store. The Q_i is sales per product or brand per store per time period. There are no accompanying data for Y and Z in the consumer framework. The data are summarized by regions of the United States, by standard metropolitan statistical areas (SMSA), and by population size. The necessary supplementary socioeconomic data can be obtained from sources such as census or population surveys.

Scanner Data

The fourth type, scanner data, is relatively new and has been rarely used in demand analysis. [See Capps (1987) for a review of recent literature on use of scanner data on demand analysis.] Scanner data are an outgrowth of new technologies to increase the efficiency of food stores by using universal product codes (UPC) and scanners at the checkout counters. The UPC and retail point of purchase scanning systems produce a wealth of new data that can be used in demand analysis. A given 11-digit UPC code represents a specific manufacturer/size/flavor/color or other related information. Point of purchase scanning can provide information on a weekly, daily, or even hourly basis. The transaction recorded by the scanner typically includes the UPC code, time of transaction, the units of the sale, and price per unit. The total value of the sale may or may not be included in the data but are easily obtained from the recorded data.

Scanner data could include a set of observations representing S individual sales of UPC item u for a specific period of time such as one day's sales in store i. Because of the vast number of original observations, a preliminary sort of the raw data would yield the following data matrix:

$$\begin{array}{l} \underline{\text{Period 1:}} \\ Q_{i11},\ P_{i11},\ Q_{i12},\ P_{i12}, \ldots, Q_{i1S_1},\ P_{i1S_1},\ C_1 \\ Q_{i21},\ P_{i21},\ Q_{i22},\ P_{i22}, \ldots, Q_{i2S_2},\ P_{i2S_2},\ C_2 \\ \cdot \quad \cdot \quad \cdot \quad \cdot \quad \cdot \quad \cdot \quad \cdot \\ \cdot \quad \cdot \quad \cdot \quad \cdot \quad \cdot \quad \cdot \quad \cdot \\ \cdot \quad \cdot \quad \cdot \quad \cdot \quad \cdot \quad \cdot \quad \cdot \\ Q_{iU1},\ P_{iU1},\ Q_{iU2},\ P_{iU2}, \ldots, Q_{iUS_u},\ P_{iUS_u},\ C_U \\ \underline{\text{Period } T\text{:}} \\ Q_{i11},\ P_{i11},\ Q_{i12},\ P_{i12}, \ldots, Q_{i1S_i},\ P_{i1S_i},\ C_1 \\ Q_{i21},\ P_{i21},\ Q_{i22},\ P_{i22}, \ldots, Q_{i2S_j},\ Pi_{2S_j},\ C_2 \\ \cdot \quad \cdot \quad \cdot \quad \cdot \quad \cdot \quad \cdot \quad \cdot \\ \cdot \quad \cdot \quad \cdot \quad \cdot \quad \cdot \quad \cdot \quad \cdot \\ \cdot \quad \cdot \quad \cdot \quad \cdot \quad \cdot \quad \cdot \quad \cdot \\ Q_{iU1},\ P_{iU1},\ Q_{iU2},\ P_{iU2}, \ldots, Q_{iUS_K}\ P_{iUS_K}\ C_U \end{array} \tag{9.5}$$

where Q_{ius} = quantity (units) of item u in transaction s in store i, P_{iut} = unit price of item u for transaction S_u in time period t for store i, and C_u = UPC code for item u.

Note that within a time period the number of transactions will be different for each UPC code, $S_1 \neq S_2 \neq \ldots S_u \neq S_i \neq S_j \neq \ldots S_k$. Note also the greater detail on food items available in scanner data. Demand analysis is possible on any one of the more than 40,000 UPC codes. Large stores will use 35,000 to 40,000 different UPC codes and smaller stores 10,000 to 15,000. There can be many UPC codes for any particular food. Since stores do not ordinarily change the price of a specific product on a daily basis, the price associated with a group of transactions (P_{iu}) will be the same. The recent advent of electronic pricing is likely to produce scanner data with more frequent price changes.

If desired, the data for an individual store can be aggregated to the city, region, or chain level. Thus, a complete machine-readable diary of sales for a given food category is available for all brands/types/and so on, as well as for complementary or supplementary products. These data can provide insights into the demand for specific products as affected by price, coupons, advertising and promotion, shelf space, and so on, variables that are not available with other data types. Generally, additional variables can be obtained from other sources within the outlet.

The major problem with scanner data is data overload. The scanner data records *every* sale of item u on a daily basis. Thus, one day's data can easily include tens of thousands of transactions ranging over more than 20,000 items in a typical retail food store. Eliminating unwanted UPC codes, sorting, aggregating across Q_i and otherwise summarizing the raw data is a nontrivial research problem.

The data set (9.5) does not include Y or Z for the individual making each purchase. For traditional demand analysis the data need to be supplemented with other data before estimating equation (9.1). Scanner data contain substantial variability due to supplemental, complementary, and competitive promotion programs, seasonal and day-of-the-week effects, nearness to payday, weather, holidays, and other outside factors influencing consumer purchase behavior. Variables to reflect such exogenous factors can be incorporated in the data from other sources. Alternatively, the maintained hypothesis implied by omitting Y and Z must be clearly examined. Some researchers argue that, for a particular store or group of stores, scanner data represent a quasi-controlled experiment. Assuming the sociodemographic variables of store patrons are constant, once the consumer is in the store the only constraint on behavior is the economic constraint of relative retail prices (Wisniewski 1984). If true, store scanner data are optimal for standard demand curve estimation. Funk et al. (1977), Jourdan (1981), and Marion and Walker (1978) have calculated price elasticities from scanner data.

Up to the present there are no publicly available scanner data sets. Most large food chains have developed their own proprietary data but they are available only through cooperative agreements with specific firms. Branson argues that scanner data combined with a consumer panel have great potential in that measurement error is reduced (Bobst et. al. 1987, chap. 8). "Behaviorscan" by Information Resources, Inc., links scanner-equipped stores to a household panel, but has been available to demand researchers on a very limited basis.

CROSS-SECTIONAL VERSUS TIME-SERIES COMPARISONS

The two most readily available data types are time-series and cross-sectional data. Each has advantages and disadvantages and imposes different assumptions on the data, the econometric model, and hence, the interpretation of the estimated parameters of (9.1). Panel data generally incorporate features of both time series and cross sections. This section discusses and contrasts selected conceptual differences and statistical problems for each data type.

CONCEPTUAL ISSUES

Clearly, the fundamental difference between time-series and cross-sectional data sets in estimating demand parameters is the time period spanned by the data. Both approaches also impose strong assumptions. First, the maintained hypothesis on how the independent variables affect demand behavior are very different. With time-series data the independent variables change over the time frame of the data set while in cross-sectional data the variables change across the consuming units within a time period. In addition, the characteristics of Q_{it} and P_{it} are assumed to be homogeneous for all t. In practice Q_{it} may vary across the data set because of subtle changes in the product. Over time new technologies lead to improved products, oftentimes with no change in P_{it}. In cross-sectional data the product purchased by high-income consumers at price P_{it} may be a different (higher) quality that the same quantity purchased at the same price by low-income consumers. Thus, the estimated demand equation (9.1) may be measuring different products across time in the time-series data or across units in the cross-sectional data.

Second, researchers find that time-series and cross-sectional data yield different estimates of a demand model. In general, the estimated income elasticities of demand using cross-sectional data are greater than those obtained using time-series data. Time-series estimates of (9.1) reflect short-run behavior while cross-sectional estimates reflect long-run equilibrium behavior.[3] The behavior homogeneity assumption implies that a relatively poor family would consume the same amount of a commodity as a rich family if its income were to become as large. However, the poor family may require time to adapt consumption levels as its income grows. Consequently, it is only in the long run that a poor family's data will reflect a consumption pattern similar to the high-income family.

Third, because of the variables available in each data type, both approaches impose additional structure on the data environment. In cross-sectional data the observations are taken over a short period of time. Hence, the consuming units are frequently assumed to face the same set of prices, even when the sample units are spread across the United States. The price effects are collapsed into the constant term, $b_0^* = b^0 + BP_{it}$. In time series the counterpart assumption of constant prices is that the characteristics of the households in the aggregates do not change over time. Average age, family size, marital status, wealth, or other sociodemographic characteristics affecting demand change slowly and are usually omitted from the Z vector. The high degree of multicollinearity of such variables is another reason for their exclusion from the statistical version of (9.1). They are usually replaced by a time trend.

Fourth, the usable degrees of freedom differ substantially. The sample size of time-series data is small in comparison to cross-sectional data. Annual data on the supply and disappearance of foods and market prices have been collected on a systematic and consistent basis since the early 1930s. In practice, most analyses use data since World War II yielding a maximum of 40 annual observations. In some cases time series are available quarterly. Thus the maximum number of data points is about 160. In contrast, one can easily find cross-sectional surveys of thousands of sample units. But the large number of individual observations requires more conditioning (Z variables) to account for interunit heterogeneity. In cross-sectional surveys the number of useful observations is inversely related to the number of independent variables in (9.1). Missing data on the independent variables, particularly income, can reduce the number of useful observations. The end result of the reduced sample may be estimates of (9.1), which are not representative of the sampled population.

Fifth, the sources of variation in Q across the observations are much greater in cross sections than in time series and the explained variation is much lower. Coefficients of multiple determination for equation (9.1) for time-series data are generally several orders of magnitude larger than for cross-sectional data, for example .9 or larger for time-series versus .3 or less for cross-sectional data. The error term in (9.1) is the result of measurement error in Q, omitted variables on the right-hand side of (9.1), and stochastic behavior of the units measured. In time-series data a part of the stochastic variation in individual behavior and measurement error in Q is averaged out in the aggregation process. As a consequence, explainable variation is a larger part of total variation. In cross-sectional data the variation in Q from observation to observation contains all three sources plus a larger random component due to individual idiosyncracies.

Sixth, the amount of detailed information in cross-sectional or microdata sets is much greater than in time-series data. In cross-sectional data values on Q_i are usually reported for detailed foods rather than the large aggregates of historical time series. For example, when using cross-sectional surveys it is easy to model the quantity demanded of particular cuts of beef, pork, or whole versus cut-up poultry. In contrast, the finest level of detail in the disappearance data published by USDA is for beef, pork, and poultry. It is possible to combine the detailed Q_i in cross-sectional data with time-series data to analyze a system of demand equations in a stagewise demand system.

Seventh, there is greater variation in prices in time-series than in cross-sectional data. In fact, cross-sectional data generally do not explicitly include price information. Until recently, researchers assumed that because of the short time span of the inquiry, all sampled units faced the same set of prices. Recent studies by Buse et al. (1985), Cox and Wohlgenant (1986), and Kokowski (1986) demonstrate that prices are important explanatory variables. Prices can be incorporated into the data set by either using published price series (Buse et. al. 1985; Kokowski 1986) or implicit prices obtained from reported expenditures by the samples unit (Cox and Wohlgenant 1986). Implicit prices in cross-sectional data can be difficult to interpret. For example, do price variations reflect real differences in supply of a homogeneous product or differences in product quality?

Eighth, it is difficult to reconcile the data from a representative cross-sectional sample with their time-series counterparts or to relate (9.1) back to farm level demand. Summing the appropriately weighted quantities in (9.3) does not equal the corresponding quantity in (9.2).

$$\sum_{i=1}^{N} W_i Q_{it} \neq Q_{.t} \qquad (9.6)$$

where W_i = the appropriate expansion weight for the jth unit, Q_{it} = quantities in (9.3), and $Q_{.t}$ = quantity in (9.2).

There are several reasons. First, time-series data from (9.2) are aggregate measures of total demand for all uses. In comparison, cross-sectional data produce an incomplete picture of demand. They do not account for all end uses. Generally, institutions such as hospitals, military and penal installations, schools, and others are not represented in consumption or expenditure surveys. Second, food eaten away from home is reported as an aggregate value rather than as quantities of beef, pork, or milk consumed away from home. It is also hazardous to try to relate cross-sectional models to farm level demand because the institutional and technological conversion factors involved in transforming a farm product into a product on the consumer's table are mainly unknown. In addition, researchers find there is a tendency of households to underreport certain types of expenditures (Gieseman 1986; U.S.BLS 1986) and overreport others (Rathje 1984). To a lesser degree, the same problem is inherent in time-series models estimated from aggregated retail disappearance data. Many primary farm products involve unknown combinations of many end uses. For example, fluid milk is converted into many distinct dairy products. The farm level quantity of highly processed products is difficult to calculate because of processing waste, nonfood uses, general waste, and other losses.

Finally, missing or incomplete data are a minor problem in time series. In cross-sectional data the problem can become a major problem. It is very likely that some observations will be missing values of dependent or independent variables. For the dependent Q_i, some surveyed units do not purchase or consume a particular food item. Because of their purchase patterns others do not report purchasing or consuming *in the survey period*. In either case, the researcher must decide whether to exclude such observations or use probit or tobit procedures that reflect such behavior. Missing information on the independent variables may be simply a nonresponse or protecting data due to confidentiality concerns, for example, top coding very large incomes. If there are missing data, observations are usually dropped from the analysis. (For details of other methods of adjusting for missing data, see Chapter 8.) Dropping observations can produce biased coefficients in (9.1), if the missing data observations are not distributed randomly among the population. It is difficult to interpret the statistical results without extraneous information on the sign and magnitude of the bias.

Statistical Problems

All types of data are susceptible to particular statistical problems. In time series, autocorrelation and multicollinearity complicate statistical computations. In cross-sectional data the problem is more likely one of heteroskedasticity of the error term, e_{it}, and zero values on Q. In panel data it is a combination of these problems. Scanner data has properties of both cross-sectional and time-series data plus omitting many important explanatory variables. The researcher is likely to encounter problems with the error term, excluded relevant variables producing biased estimates on the included variables, and difficulty in adding variables to reflect nonprice or promotional effects (services, cleanliness, reputation, coupon use). There has been so little experience with scanner data in demand analysis that there is no literature on statistically peculiar problems. Nevertheless, they surely exist and will be appearing in the literature in the near future.

Time-series data exhibit serial correlation of the residual and correlation among the independent variables. The model variables tend to change in a similar pattern from one time period to the next because the general level of economic activity affects all the variables in the model. Since the error term reflects conditions not explicitly accounted for in the model (such as omitted variables), first-order autocorrelation in the residuals is also usually encountered.

Time-series data also exhibit multicollinearity for the same reasons, that is, the tendency of the independent variables to move together. For example, income and prices exhibit the same cycles and trends over time. Thus, there may be too little independent variation to accurately calculate the effect of the independent variables on the dependent, ceteris paribus. All four types of data can suffer from small independent variation in the independent variables making it difficult to estimate particular parameters accurately. This is usually the most serious in time-series data, although prices in cross-sectional data also exhibit a high degree of multicollinearity.

Another problem with time-series, cross-sectional, and panel data is operationalizing known structural change. The change may be because of shifts in consumer behavior or because of changes in the data generation process. If the structure has changed, then two different date points reflect different behavior by the same consumers. For example, following World War II consumers reacted differently to price and income change than they did before Pearl Harbor. Similarly, there is research evidence that the consumer's response to price changes following the oil crisis of the early 1970s was different than before the change but exactly where and how the behavioral change took place is not clear. Recent public health information on diet may have changed the consumer's attitudes towards foods such as poultry, fish, dairy products, and red meat. Finding the appropriate variable to reflect the new information on diet and health to incorporate into equation (9.1) is difficult.

A change in data conceptualization by those who collect and publish the data is a second type of structural change. That is, the data may have been revised or the method of calculation changed. In time-series data the coverage of a particular subitem may have changed or the procedure used to calculate it improved. In 1982, ten fresh vegetables and several fruits were discontinued due to cutbacks in the USDA budget for collection production data. Nelson and Duewer (1986) describe the procedure for estimating loss, waste, and nonmarket consumption in calculating beef disappearance. The net result is a change in the "quality" of the disappearance data. In cross-sectional data similar changes in methodology have taken place. For example, since 1980 when data is missing or inconsistent in the consumer expenditure surveys, the BLS imputes values to those variables. It also allocates expenditures when reported values are not specific. For example, if a single amount is reported for a breakfast of bacon and eggs, that value would be allocated partly to bacon and partly to eggs. The problem is that improved conceptualization is reflected in the published data but the exact nature of the revision may not have been widely published. A change in data coverage is a third possibility. For example, between 1981 and 1983 the BLS consumer expenditure panels did not sample rural areas.

Heteroskedastic error terms are the major statistical problem of cross-sectional data. The heteroskedasticity arises out of the fact that the errors are related to household characteristics in the Z matrix. For example, low-income households have less income to allocate across foods than high-income households. As a consequence, there is more choice and variation in Q_i for high-income households. Since income is highly correlated with other sociodemographic measurements, the error variance is usually related to many sociodemographic characteristics of the consuming unit. Again, the most appropriate correction is unresolved.

Multicollinearity among particular subsets of variables such as age, income, and education can also be a problem in cross-sectional data. Multicollinearity increases the difficulty of testing hypotheses on the parameters. Income is generally highly correlated with age and education, making it difficult to obtain statistically reliable results for selective effects. The more socioeconomic variables included in Z the more complex is the multicollinearity. If prices are incorporated into a cross-sectional analysis they also are highly intercorrelated.

Errors in measurement can also be severe in cross-sectional data sets. The variables are usually based on the interviewee's recall or recording ability. Research indicates that many variables in microdata sets contain substantial measurement error. But most statistical estimation procedures assume the independent variables are measured without error. These errors can substantially affect the estimates. Duncan and Hill (1984) illustrate the magnitude of the problem using respondents' recall of labor earnings. Altonji and Siow (1987) show the effect on the consumption function of measurement error in income.

Zero values for the dependent variable is another statistical problem in cross-sectional data. For highly disaggregated demand analysis a certain percentage of the respondents will report a quantity of zero. The zero can be interpreted as *not ever* consuming that food item or not consuming or purchasing in the period of inquiry (for example, yesterday, last week, or last month). In either case, the data are said to be censored at zero. The econometric implications of a censored data set are beyond the scope of this chapter but require estimation procedures other than OLS (see Maddala 1983). There is an alternative approach to circumvent zero values on the dependent variable. This approach subsets the data into groups of similar types of units such as all single-person households, all two-person households aged 25 to 35, and/or all households aged 36 to 55. The procedure uses mean expenditures and income for each socioeconomic cell to estimate the parameters of (9.1). Using cell averages or aggregates requires assuming macrobehavior and average macrobehavior are equal. Using group averages also introduces a weighting problem. Each cell likely contains different numbers of units and thus different amounts of information are included in the aggregates of each cell. The estimation procedure must account for those differences if the estimates are to have desirable statistical properties. (See Chapter 8 for details.)

Panel data can exhibit both autocorrelation and heteroskedasticity. The error variance of the sample of observations may be different across both time and the units. For example, the error variance of an individual for one time period may be about the same as for the next period. Thus, the error variance of e_{it} across time will be correlated. Simultaneously, the error variance of an individual observation may be related to the income level of that unit. If the error variance of e_{it} is heteroskedastic and autocorrelated, classical least squares is not appropriate to estimate the parameters of (9.1).

TIME-SERIES DATA SETS

Almost all time-series data used in demand analysis of U.S. agricultural commodities are collected and published by USDA and BLS. This is a by-product of the agent's responsibilities and, as such, the "match" to the economic concepts implied by the theory of (9.1) is imperfect. The time-series counterpart of Q_i is the USDA disappearance data. For prices, P_i, there are several alternative price series.

DISAPPEARANCE DATA

Disappearance data are the most comprehensive measure of the consumption and utilization of domestically produced food in the United States. They measure domestic market supply in a given time period. The time period for most commodities reported is annual, although a few commodities are reported for shorter periods. The length of the time series varies by commodity, with the longest being since 1909. Disappearance data include more than 200 individual agricultural commodities.

The USDA regularly produces estimates of supplies of domestically produced agricultural products. Food disappearance quantities are calculated as residuals from total estimated production adjusted for stock changes, imports and exports, nonfood and nonmarket uses (see USDA 1988). The residual is converted from a primary commodity weight (e.g., carcass weight, fluid milk equivalent) to the retail weight equivalent. Retail weights of farm level commodities are difficult to calculate because many of the institutional and technological conversion factors are not precisely known. Processing waste, nonfood uses of parts of the farm product, other wastes, and losses in marketing and storage are examples of the factors that must be subtracted from farm level production to obtain retail weights (see Nelson and Duewer 1986). The estimated retail weight is divided by estimated civilian population to obtain annual per capita disappearance or consumption.

The primary use of disappearance data is to provide a description of the flow of U.S. produced agricultural commodities into end uses. Since they measure total disappearance they include both at-home and away-from-home consumption. If it is assumed that the commodity market in question is in equilibrium over the observational period, the disappearance is conceptually equivalent to the quantity clearing the market at prevailing prices. The data are published regularly in USDA's *Agricultural Statistics* and in its annual statistical bulletin, *Food Consumption, Prices, and Expenditures* (see USDA 1987).

PRICES

Prices are published at several levels of the marketing system. It is important to match the disappearance data with the appropriate prices. State and national farm level commodity prices as of the fifteenth of the month are reported in *Agricultural Prices*. They are based on a survey of commodity dealers' prices paid to farmers. The prices are representative of the average quality sold by farmers at the first level of sale. The Agricultural Marketing Service also reports prices by grade and variety on designated central wholesale markets in commodity publications. Wholesale prices are also collected for many products by the Bureau of Labor Statistics. Individual product prices and product group prices are reported monthly in the *Survey of Current Business*.

Consumer prices are gathered and published on both a monthly and annual basis by the Bureau of Labor Statistics in the Consumer Price Index (CPI). For analyzing the demand for food most researchers use the consumer price index for all urban consumers (CPI-U) or the CPI for wage earners (CPI-W). The CPI-W has been published for more than 50 years. It covers urban hourly wage earners and clerical workers in the civilian noninstitutional population. The CPI-U has been published since 1978. It is a broader index in that it covers all residents in urban places. It is widely used as a general measure of inflation to show changes in the real purchasing power of the income variable in (9.1). The ratio of Y to CPI measures changes in real income. The CPI-U can be viewed as an approximation to the cost of a "constant standard of living." It is imperfect in that it does not allow for substitution as a result of change in relative prices. Furthermore, it covers only consumption expenditures and does not include government services and other "nonmarket" activities such as pollution control or product safety. They are also part of the cost of a standard of living (see Cagan and Moore 1981 for more detail).

The CPI-U is a statistical measure of price changes for a set of goods and services included in a representative "market basket." It measures retail prices for about 80 percent of the U.S. civilian population. Currently, it is published monthly for 28 major cities, for regions and various sizes of urban areas, and a national average. The city and regional CPI measures changes in specific cities, urban areas, or regions over time but *cannot be used as a measure of price differences among locations*. (See Blanciforti and Parlett 1987.)

The CPI includes prices for 16 major food groups including food at home, cereals and bakery products, beef and veal, pork, other meats, poultry, fish and seafood, eggs, dairy products, fresh fruits and vegetables, processed fruits and vegetables, sugar and sweets, fats and oils, nonalcoholic beverages, other prepared food, and food away from home. Prices are also published for a number of specific subitems within each of the food groups. For example, in addition to a price series for beef and veal, prices are collected and published for ground beef, chuck roast, round roast, sirloin steak, and other beef and veal. The two series also include prices for all the nonfood components of the market basket.

The BLS also publishes monthly producer prices for 15 different food groups. The producer price indexes reflect price changes at various levels in the production and marketing channels.

A major problem with matching price data with disappearance data is that there is not a close correspondence between the quantity in the USDA disappearance data and reported prices. The quantities in the disappearance data are "retail weight equivalents" based upon estimates of production. They are indirect estimates of the actual consumption at the retail level while the price data are for actual quantities at a specific level in the market channel. The quantity estimates total volume, that is, beef, pork, or vegetables moving into consumption, while the reported retail price is for a specific item such as the weighted retail price of beef or a price for sirloin steak, ham, or frozen vegetables, while agricultural prices reflect the first level of sale beyond the farm gate such as a livestock auction, fruit or vegetable packing plant, or other wholesale market.

INCOME

Income is included in (9.1) to measure changes in the purchasing power of consumers. A frequently used income series is per capita disposable income from the national accounts. It is published annually and quarterly by the U.S. Department of Commerce in *Survey of Current Business*. If a more detailed breakdown of income is desired, the Bureau of the Census annually estimates the per capita income of families and individuals. Specific estimates are provided for different regions, ethnic groups, occupation, household types, and ages on a national basis.

MAJOR CROSS-SECTIONAL DATA SETS

The following describes the design of and information contained in recent surveys of U.S. households useful for demand analysis. They include data on food consumption or expenditures. The surveys described are (1) consumer expenditure surveys (CES), (2) nationwide food consumption surveys (NFCS), and (3) national health and nutritional examination surveys (NHANES). These microdata sets are large, ranging from several thousand to tens of thousands of observations. Information on how to obtain machine-readable data can be obtained from the National Technical Information Service of the U.S. Department of Commerce or the Division of Consumer Expenditure Surveys of the BLS.

BLS CONSUMER EXPENDITURE SURVEYS (CES/I, CES/D)

The consumer expenditure surveys are conducted by the Bureau of Labor Statistics. It is responsible for publishing the nation's price indexes, and the periodic surveys provide the basic information used in calculating the weights for those indexes. The surveys collect detailed information on the income, expenditures, and financial status of a representative sample of U.S. households. There have been surveys in 1888–91, 1901, 1917–19, 1933–36, 1941–42, 1950, 1960–61, 1972–73, and on a continuous basis since 1980 (see Carlson 1974). The first survey obtained cost-of-living data from U.S. wage earners. Subsequent surveys were of wage earners and salaried workers in urban areas. Starting with the 1960–61 survey both the scope and sample have been expanded. The sample is now for the entire noninstitutional population of the United States. In addition to the data required for expenditure weights, the 1960–61 survey collected information on household sociodemographic characteristics, assets and liabilities, and income. The 1960–61 survey used a seven-day recall for food expenditures and a longer recall period for other expenditures. The recall methodology provided little detail for food expenditures and the accuracy of recalling past purchases was questioned. As a result, the methodology of the succeeding 1972–73 and later surveys was changed.

Starting with 1972–73, BLS used two separate surveys, each with its own questionnaire and sample. The rationale was that the recall of expenditures varied with the cost and importance of the item. Information on large, more easily recalled expenditures such as housing, utilities, clothing, furniture, appliances, health care, recreation, transportation, and educational expenses were collected in a series of five quarterly interviews covering a 15-month period. This survey is commonly referred to as the *Interview* (CES/I). The CES/I also obtained global estimates for more frequently purchased items such as food, beverages, and utilities. The CES/I was designed to obtain detailed information on 60 to 70 percent of a household's expenditures and aggregate information on another 20 to 25 percent. Details on these global aggregates of frequent purchases and other remaining expenditures were obtained in the second concurrent survey, the *Diary* survey (CES/D).

The CES/D collected information via a daily diary on the 30 to 40 percent of the most frequent expenditures. Participating households were requested to keep a detailed diary of purchases over two consecutive one-week periods. The expenditure components covered by the diary are food, household supplies, personal care products, and nonprescription drugs. The BLS publishes information for more than 50 different expenditures from the interview and for 20 different foods from the diary. In addition, BLS publishes integrated data from the interview and diary (see U.S. Bureau of Labor Statistics 1978).

Up until 1960–61 the data were only available from BLS as published in its statistical bulletins. The 1960–61 survey marked the first time the BLS released the microdata in machine-readable form (magnetic tape). Data from all BLS expenditure surveys since 1960–61 are available in machine-readable format. The level of expenditure detail and socioeconomic information in these tapes can be overwhelming. The detail tapes contain information on thousands of different expenditures and on the characteristics of each member of the sampled unit.

Sample Design. The sample design of surveys prior to 1960–61 varied from survey to survey. Generally, surveys concentrated on urban areas. In some surveys, the CES was augmented by the USDA's survey of farm families. The 1960–61 survey sampled both rural and urban areas of the Unites States. It included 13,728 usable schedules. The 1972–73 and later surveys sampled the noninstitutional population of the United States and included all urban, rural, farm, and nonfarm households. The 1972–73 interview sample size was 19,975 consumer units and the diary sample was 23,186.

Data. In addition to expenditure data, all surveys collected information on the household's socioeconomic characteristics. The more recent surveys contain more of this information than the earlier surveys. Both the interview and the diary collect detailed information on the socioeconomic characteristics of the consuming unit, for example, age, sex, race, education, and marital status of each member; location, housing, and tenure (owner/renter); and occupational characteristics of head and spouse.[4] Both instruments also collect details on the consumer unit's income—including in-kind income and a set of questions on food stamps. Because of differences in collection methods, the income reference period in the CES/I is not the same as for the CES/D. The diary reference period is the previous 12 months while the survey reference period is the calendar year. As a further difference, diary incomes cover varying 12-month reference periods.

In the (CES/I) each consuming unit (CU) is queried in detail about all periodic expenses. The interview expenditures cover clothing, utilities, furniture, large and small appliances, real estate, motor vehicles, information on out-of-town trips and vacations, taxes, repairs, insurance, professional services, and contributions. In the available detailed tapes, expenditures for each consumer unit are recorded for more than 3,500 different items. In addition, global estimates of expenditures on food at home, food away from home, alcoholic beverages, drugs, medicines, and personal care are also obtained. These latter items are covered in detail in the CES/D.

The interview (CES/I) also includes a detailed inventory of major durable goods such as automobiles, stoves, and dishwashers; characteristics of those durables such as whether the automobile had air-conditioning or a standard transmission, the refrigerator is frost free, and so on; an inventory of minor appliances, and assets and liabilities of the CU. The information on sources of income is also more detailed than in the CES/D.

The major components of the diary (CES/D) include food, household supplies, personal care products, nonprescription drugs, and housekeeping and garden supplies. Within each broad category a very detailed set of items is available. For each individual or consuming unit, the records can include several thousand different items such as wheat flour, T-bone steak, 2 percent milk, canned vegetables, vacuum bags, new appliances, used appliances, books, magazines, power tools, water softening service, motor oil, and cooling system repairs. In all, more than 1,700 different expenditure items are recorded with more than 1,200 for food. The recorded expenditure may be for either one or two weeks.

The BLS data surveys record expenditures, but there is no explicit price information. This is a major limitation for the researcher interested in demand analysis, since the assumption of constant prices across the time frame of the more recent surveys is questionable. For food, there has been some work on estimating price information from the quantities and total expenditure data, *when both are available* (see Capps and Havlicek 1981). Obtaining food quantities from BLS tapes requires a substantial amount of preprocessing and data cleaning. [See Capps et al. (1981) for information on obtaining implicit prices from the 1972–73 BLS/D.]

USDA Nationwide Food Consumption Surveys (NFCS)

The USDA has conducted national surveys of food consumption about every 10 years. These surveys provide information on the quality of diets and the amount of money spend on food by U.S. households. Surveys were conducted in 1936–37, 1942 (spring only), 1948 (urban areas only), 1955 (spring only), 1965–66 and 1977–78. The first four surveys included detailed information on household food use. Later surveys were expanded to include information on food consumed away from home and to seasons other than the spring quarter. In 1965–66, data were collected for all four seasons of the year and information on the food intake of each household member in the spring quarter of 1965 was added to the survey.[5] In the 1977–78 survey, food intake of individuals was obtained for each season of the year. Work is currently proceeding on a 1987–88 survey.

Sample Design. The NFCS uses a national sample representing the U.S. population in the coterminous 48 states. At times there has been a special emphasis on selected subpopulations. Details on those surveys available in machine-readable form follow.

The 1955 survey was a national self-weighting probability sample from the 48 states. It included 4,556 household supplemented by 1,504 farm operator households. An eligible household was one in which at least one member had 10 or more meals from household food supplies. Data were collected in the spring quarter of 1955 (April, May, and June).

The 1965–66 NFCS sample was selected to represent all areas of the United States, with the exception of Alaska and Hawaii. Similar to 1955, the sample design was a national self-weighting basic sample plus a supplementary farm sample. The final sample included 15,101 households of one or more members. The sample excluded people living in group quarters such as rooming houses, hospitals and prisons, and those households in which no member ate at least ten meals from the home food supply during the seven days preceding the personal interview. The interviews took place between April 1965 and April 1966.

The 1977–78 survey sample is representative of households in the 48 coterminous states. The survey was conducted from April 1977 through March 1978 and included several special supplementary samples. The basic sample included 15,000 households and 34,000 individuals. Supplemental samples included an Alaska sample of 1,100 urban households (2,400 individuals) sampled in the winter quarter of 1978; a Hawaii sample of 1,250 households (3,100 individuals) sampled in the winter of 1978; a Puerto Rico sample of 3,100 households (900 individuals) sampled in the last six months of 1977; an elderly sample of 5,000 households with at least one member 65 years old or older (7,500 individuals); and a low-income sample of 4,700 households eligible for food stamps (13,000 individuals) sampled between November 1977 and March 1978. There was also a special sample in the spring of 1977 called a bridging sample. The purpose of the bridging sample was to provide comparable data for evaluating the effect of changes in methodology between 1965 and 1977–78. Although the methodology for the 1977–78 NFCS was quite similar to the 1965 survey, there were some differences (see USDA 1982). The results indicated that the methodological changes had no effect on the tabulated results. A comparable survey for low-income households was also conducted in 1979–80.

Data. The NFCS is designed to measure home food consumption of members of the sample household, not food expenditures. Generally, the NFCS contains varying levels of detail on the sampled households. All surveys include information on socioeconomic characteristics, income, family size and composition, the quantity and value of food consumed from household food supplies, and the sources of the food (purchased, gift, or home produced). Each survey also had unique sets of questions not found in other surveys. In general, the more recent surveys provide more socioeconomic detail than earlier surveys and also more detail on the food intake of individuals in the sample.

The 1977–78 Nationwide Food Consumption Survey provides two levels of detailed information on the types and amounts of food used at home. The first level is a recall of food used by the entire household *at home* in the previous seven days *from home food supplies*.[6] Each food item included the quantity, form (fresh, frozen, canned, or dried), source (purchased, home produced, gift, or pay), and price (if purchased). Foods consumed away from home in restaurants, schools, and cafeterias were not included, although aggregate expenses for food bought and eaten away from home were recorded.

The second level of detail was an individual intake record for each household member. In the spring of 1977 all individuals in all households were asked to provide food intake information. In the other three seasons all individuals under the age of 19 years and 50 percent of those 19 and older provided food intake information. Individual intake information was also available for Alaska, Hawaii, Puerto Rico, and the low-income survey. These data can be used to study the demand for nutrients (Basiotis et al. 1983).

The intake records in the NFCS detail the kinds and quantities of food consumed both at home and away from home. They were based on a 24-hour recall plus a two-day food diary. From the households participating in the survey, 30,770 individuals completed at least one day's food intake and 28,030 completed data for three days. For each food, the quantity eaten, form in which eaten, source of the food, and eating occasion were recorded. From those data USDA estimated the nutrient content of the household's food consumption and the nutrient intake of individual household members. This has now evolved into the Continuing Survey of Food Intake of Individuals (CSFII) described in detail in the next section. From the NFCS more than 4,500 individual food items and 14 nutrients plus calories can be studied. The 1987–88 NFCS has been expanded to include 27 nutrients plus energy. Since the recorded information includes both quantity and value, price per unit can be inferred.

The sociodemographic characteristics available include information on income, household size and composition, race, region, urbanization, general food shopping practices, number of meals and snacks from household food supplies, a description of the dwelling, educational level, and employment status for the heads of the household, participation in food programs, and the sex and age of each household member.

The National Health and Nutrition Examination Survey (NHANES)

The National Center for Health Statistics (NCHS) of the Department of Health and Human Services has been collecting statistics on a wide range of health conditions for more than 20 years. The objectives of the NHANESs are to measure the health and nutrition status of the U.S. population, to monitor and describe health and nutritional conditions, and to provide information on the prevalence of diseases. In addition they provide normative measures of U.S. population characteristics such as weight and height over time. Although the primary purpose of the NHANES is to collect data on the health conditions of our population, it also contains information on individual food intake. In some respects it is an alternative source of food intake data similar to the USDA/NFCS. (For a discussion of similarities see Coordinating Committee on Evaluation of Food Consumption Surveys [1984] and Swan [1983].) The NHANES emphasis is health and nutrition; consequently the economic details on income and food expenditure information are very limited in comparison to the NFCS.

The National Center for Health Statistics conducted its first nationwide health examination survey in 1960–62 with adults 18 to 74 years old. During the 1960s two additional surveys were conducted with children 6 to 11 years old and adolescents 12 to 17 years old. The second survey was initiated in 1971. It included new information on food consumption and on the nutritional status of the individual. Nutritional status was assessed through a complete medical history, questions on diet, anthropometric measurements, a physical examination, and other medical tests. This survey, called the National Health and Nutrition Examination Survey (NHANES-I), was carried out from 1971 to 1974. It sampled the U.S. population 1 to 74 years old. NHANES-II, conducted between 1976 and 1980, extended the surveyed population to infants 6 months of age or older.

In 1982–84 NCHS conducted the Hispanic Health and Nutrition Examination Survey (HHANES). The survey sample was limited to the area of the United States in which there was a large Hispanic population. It surveyed persons of Mexican-American, Puerto Rican, and Cuban ancestry living in the southwestern United States, the New York City area, and Dade County, Florida. NCHS conducted NHANES-III in 1988.

Sample Design. In contrast to BLS and USDA surveys, the NHANESs are samples of the U.S. population rather than samples of households or consuming units. NHANES-I is a probability sample of the U.S. civilian, noninstitutionalized population of all 50 states aged 1 to 74 years. The sample was stratified by broad geographical regions and by socioeconomic characteristics within regions. It included very detailed health examinations for the population between 25 and 74 years of age. In NHANES-II the sample was expanded to include persons between 6 months and 74 years. Because preschool children (6 months to 5 years), the aged (60 to 74 years old), and the poor (persons below the poverty line as defined by the U.S. Census Bureau) were assumed to have more malnutrition problems, they were oversampled. NHANES-I included about 20,750 individuals and NHANES-II approximately 21,000 individuals.

Data. For each participant, NHANES records data on family relationships, sex, age, and race of family members; housing characteristics; occupation, income, and educational level of each household member; and participation in the food stamp, the school breakfast, and the school lunch programs. Dietary information includes a recall of food intake for the preceding 24 hours, usual food consumption during the preceding three months, diets, medications, and vitamin and mineral supplements. Medical data include anthropometric measurements, allergies, and the results of a series of clinical assays of blood and urine.

Food consumption includes estimated portion sizes of all foods and beverages consumed in the previous 24 hours, the time of day the food was eaten, and its source (home, school, restaurant, or other). For each food the quantity consumed and the nutrient content, fats, vitamins, and minerals are also recorded. The consumption data are reported by food (about 4,800 food items). The frequency of consuming 18 groups of foods in terms of never, daily, weekly, or less than once weekly; the type and frequency of alcohol consumption; and the use of salt, special diets, and commonly prescribed medications are also obtained. The NHANES includes no explicit prices or sufficient data to calculate implicit prices.

PANEL DATA

The following is a description of the design, operation, and data of those panel data sets considered useful for analyzing the demand of U.S. households or individuals for food. It includes data on either food consumption or expenditures by households or food intake by individuals. Generally, the food intake data can be aggregated to a household level, if the analysis requires. The data sets are

1. Continuing Survey of Food Intake by Individuals (CSFII)
2. BLS continuing consumer expenditure surveys (CES/I, CES/D)
3. University of Michigan Income Dynamics Panel
4. Longitudinal Retirement History Study (LRHS)
5. Special university panels
6. Private panels

CONTINUING SURVEY OF FOOD INTAKE BY INDIVIDUALS (CSFII)

The CSFII is a yearly nationwide survey of the food and nutritional intake of selected groups of individuals in the coterminous 48 states. The first was conducted in 1985 and is one component of the Federal National Nutrition Monitoring System. The CSFII provides annual updates and details of diet adequacy of selected subgroups of the population and timely indications of dietary changes in the sample populations. It is designed to complement the larger NFCS conducted every 10 years, the most recent being in 1977–78 and 1987–88.[7]

The primary focus of the first CSFII was on households containing women 19 to 50 years of age and their children 1 to 5 years of age. This group is referred to as the "core monitoring group." It was selected because previous surveys show that women of childbearing age and their young children are more likely than other groups to have diets deficient in certain nutrients. Other age and sex groups are also included but with less frequency.

Sample Design. The CSFII is a stratified sample of households in the 48 states. The sampling procedures reflect geographic location, degree of urbanization, and the socioeconomic characteristics of areas. The sampling and screening procedures produced three separate samples: (1) women 19–50 years of age and their children 1–5 years old (the core monitoring group), (2) a sample comparable to (1) of low-income women and their children, and (3) men 19–50 years old. The final sample of the 1985 CSFII included 1,342 households, providing information on 1,503 women, 550 children in the core monitoring group, a comparable low-income sample of 2,120 women and 1,314 children, and 658 men in the third sample.

The data are a pseudopanel in that each succeeding year a new panel of subjects in the same age-sex groups is selected. The survey design permits adding supplementary samples of other age-sex categories as funds and interest dictate.

Data. The CSFII collects data on the previous day's food intake of individuals. In 1985 men were surveyed once during the year while women and children are surveyed on six separate days over a one-year period. The food intake information is converted into 27 dietary components plus energy. In contrast, the 1977–78 NFCS individual intake data were converted into 14 nutrients plus energy.

Information includes all food eaten by the individual either at home or away from home, the time of day it was consumed, the use of salt and fat, and the form in which the food was brought into the household (commercially frozen, canned, bottled, or dried). Each woman in the sample also provided information on age, race, whether pregnant or nursing, employment status, occupation, education, and use of special diets, vitamins, and mineral supplements. Information on household characteristics includes the previous year's before-tax income; participation in food programs; the male head of household's age, education, occupation, and employment status; household size; tenancy; usual amount spent on food; and each household member's sex, age, and relationship to the female head of the household. Since the unit of observation is the individual, there is no price or expenditure information. Individual data can be related to the household of which they are members.

BLS Continuing Consumer Expenditure Surveys

In October 1979 BLS began a continuing survey of consumer expenditures. It continues up to the present and can be classified as a pseudopanel in that the sampled units are replaced periodically. It is patterned after the 1972–73 survey and consists of two separate samples and questionnaires. One is a diary survey in which the sampled unit is asked to complete a diary of expenditures for two consecutive one-week periods. The other is a quarterly interview in which each sampled unit is visited by an interviewer once each three months over a 12-month period. The diary obtains data on frequently purchased items including food, personal care items, and household operations. The interview obtains expenditures on less frequently purchased items such as major appliances, autos, rent, and insurance premiums. The continuing CES is emerging as a very detailed time series of cross sections. By 1988, six years of data will be available for analysis. In addition BLS plans to publish integrated data from the interview and diary.

Sample Design. The continuous interview visits 5,000 consumer units every three months over a 15-month period and then replaces the unit with a new household. The data from the last four visits is included in the data set. The diary sample is of 5,000 different consumer units visited twice in two consecutive one-week periods. Because of budget cuts, rural consumer units were dropped in the fourth quarter of the 1981 survey and then resumed in the fourth quarter of 1983. Thus, published data for 1980–81 is only for urban households although data tapes do contain rural observations for 1980 and part of 1981.

Data. The data in the continuing surveys are very similar to those of the earlier BLS diary and interview. The reader is directed to the associated earlier section for detail.

UNIVERSITY OF MICHIGAN INCOME DYNAMICS PANEL

Since 1968 the survey research center of the Institute for Social Research at the University of Michigan has maintained a panel of almost 5,000 families. The purpose of the panel is to study the determinants of family income and its changes. The long-term study hopes to gain a better understanding of the dynamics of household economic behavior. Currently, there are almost 20 years of data available on the same families and their descendants. All new families formed by members of the original families are added to the sample. Each spring, heads of the families are interviewed about attitudes, economic status, and economic behavior.

Sample Design. The sample is a representative cross section of the United States for about 3,000 families plus a subsample of 1,900 low-income families. The combined sample can be weighted to be representative of the total U.S. population. The current sample size includes over 6,500 families because of the procedure of adding new families that contain adult members of the original families.

Data. The data available from the income dynamics panel fall into three categories: questions on economic status, economic behavior, and attitudes. The data include information on income, employment, housing, auto ownership, food expenditures, transportation, education, marital status, family composition and background, and measurements on attitudes. Additional questions on topics of special interest are added from time to time. Household food stamp participation is also available. The survey collects little information on food expenditures outside of categorical data on food consumed at home and food consumed away from home. There is no detail on food expenditures available on a regular basis and no price information in the data set. Nevertheless, the data have been used for estimating aggregate food demand incorporating regional food prices for BLS/CPI (Benus et al. 1976).

Longitudinal Retirement History Study (LRHS)

The Social Security Administration initiated the Longitudinal Retirement History Study (LRHS) in 1969 (Ireland 1972). Its objective was to gather information useful in studying the process of retirement and changes in such households over time. It investigated the reasons for early retirement and the changes in the economic and social characteristics of older persons as they approach and enter into retirement. In order to explore the factors most affecting the timing of retirement and the changes in life-style and living standards the survey design followed a sample of individuals aged 58 to 63 in 1969 for 10 years. The individuals were reinterviewed in odd numbered years through 1979. It was a true panel.

Sample. The LRHS sample was drawn to represent all persons born between 1905 and 1911 in the 50 states. The study was designed to begin with preretirees who were then followed into retirement until ages 68–73. The initial sample included 11,153 people. The study included men aged 58–63 and women of the same age in households without husbands. It did not include women who were living with husbands when the sample was selected because to such women the concept of retirement usually meant their husband's retirement, not their own.

Data. Information collected in the LRHS included basic demographic data; work history; health and living arrangements; leisure activities; spouse's work history; financial resources and assets; expenditures; family composition; household, family, and social activities; and retirement plans. The LRHS obtained information on the respondents, expenditures for the major budget items of food, shelter, transportation, and medical care. Expenditure data are also available for personal care, entertainment, dues, gifts, and contributions (see Social Security Administration 1976).

The LRHS developed a limited data base. It was a study of the living patterns of older American households but did not include an exhaustive list of their expenditures. Consequently, it is useful for limited demand analyses that focus on the elderly households and do not require complete budget information. It also does not permit calculating or otherwise obtaining prices from the data set.

SPECIAL UNIVERSITY PANELS

There are several consumer panels operated by academic institutions that contain very comprehensive information on food purchases (see Raunikar and Huang 1987, 41–42). These panels operated for various periods of time. More information can be obtained from the university that operated the panel. The University of Georgia operated a consumer panel in Griffin, Georgia, between 1974 and 1981 and in Atlanta, Georgia, between 1956 and 1962 (Purcell and Raunikar 1971). Detailed information on the purchase and use of food was also collected by Michigan State University in East Lansing, Michigan (Quackenbush and Shaffer 1960), and by North Carolina State University in Raleigh, North Carolina, during the late 1950s and early 1960s. The Puerto Rico Agricultural Experiment Station also conducted a continuous household panel for several years beginning in October 1978.

Private Panels. There are a number of panels of consumers and retail outlets operated by marketing research firms. The information from the panels is summarized and sold as a proprietary product. The major panels are described below. For more details and addresses of the firms operating the panels see Forker et al. (1987).

Chain Restaurant Eating Out Share Trend (CREST). CREST is operated by the National Purchase Diary Group (NPD). It is based on data collected from a continuing panel of 10,000 families and 2,800 singles. All 12,800 households report on a quarterly basis producing data that can be aggregated into away-from-home purchase for eight weeks of the year.

Information is recorded for each meal occasion on the type of eating establishment, the meal eaten, day of the week, total cost, tip, and what foods were eaten for each family member. Socioeconomic and demographic information on the sample family includes its size and composition, race, income, education, hours of work for husband and wife, location of residence, and home ownership status.

CREST is probably the best current source of data on food consumed away from home. Researchers other than industry have purchased parts of the data for demand analysis to augment data from other areas (Folwell and Baritelle 1978). Since the survey is generated by the private sector, the data are likely to be costly.

The NPD group also operates a number of other special purpose panels that could be useful for particular demand studies. The Packaged Goods Diary Panel provides information from two national panels of 6,500 households and 40 local panels of 1,000 households each. Panelists record their purchases over a month for all items that are listed in the diary. The items vary from period to period depending upon client interest. It also operates special purpose National Food Consumption Panels. These panels provide in-depth analysis of at-home food preparation and consumption patterns.

MRCA Panel. The MRCA panel is one of the oldest nationwide household panels. It is operated by the Market Research Corporation of America. It is a nationwide sample of households that record purchases of *selected* items in a weekly diary. The sample includes 7,500 households. The panel has been selected to represent different demographic and geographic classifications. The data can be projected to the national level by region and demographic groups. The data set is generally not available for public research but has been used for special demand studies (Boehm and Babb 1975a, 1975b).

Products covered include food and beverages, health, beauty, and personal products. The specific products vary since they are restricted to items for which firms are willing to buy data. The purchase information in the diaries includes UPC codes, where purchased, and characteristics of the product such as brand, type, and flavor. The data can be related to household socioeconomic information.

Mail Diary Panel (MDP). MDP reports information on a limited number of grocery, drugstore, and household items through monthly mail diaries from 5,000 households. The sample can be weighted to the national level. The products included in the diaries are highly variable and depend on MDP client's products.

Market Sales and Purchase Data

A number of private firms provide continuous data on a sample of retail stores. They track the sales or movement of food store products at the wholesale and retail levels. The information is sold to clients interested in the movement of particular products in specific market areas.

Scantrack US is a product of the A. C. Nielsen Company. It records sales of selected products in a sample of 1,600 U.S. retail outlets using scanners. It also provides similar data in 23 other countries. The sample of stores is stratified by store type, size, service type, ownership, and income level of the area it serves. All items that have UPC codes are included in the data. The data are available on a weekly basis. The data can be combined with outside factors affecting sales such as weather conditions, unemployment rates, local pay schedules and government assistance, and the demographic characteristics of the market in which the sample outlet is located. It has been used for demand analysis on specific products such as orange and grapefruit juice (Brown and Lee 1986).

SAMI (Selling Areas Marketing Incorporated) is another private system that collects product movement data at the processor and wholesale level for most food chains. It has collected data since 1966 and is used to estimate volume movement across the United States. SAMI reports shipments from warehouses to retail stores for most major food chains in most standard metropolitan statistical areas (SMSAs). The data base covers 468 categories of goods incorporating more than 220,000 specific items. Manufactured dairy products, juices, and other foods that move through warehouses as part of the wholesaling process are included. The SAMI data do not include information on fresh meats, fruits, or vegetables since they do not pass through warehouses. There are additional private firms selling this type of data but most of them are not very useful for demand analysis. They do not include most of the conditioning variables required to reliably estimate demand equations like (9.1).

Behaviorscan and Infoscan are produced by Information Resources, Inc. (IRI). The panels are designed to track consumption in response to promotional, pricing, and other merchandising campaigns. The Behaviorscan households are linked to cable television networks for testing promotional programs in a set of sample communities. Each participating household has a computerized ID card, which is used to automatically record purchases on store scanners in those communities. Behaviorscan relates household characteristics to the scanner records of the sample household's purchases. Ten communities have been participating in the Behaviorscan system for up to eight years. There are 3,000 to 3,500 households in each of the communities. Infoscan tracks consumer purchases of UPC-coded products sold in supermarkets. The unit of observation is the supermarket. The sample encompasses 2,000 stores in 53 different market areas.

For both Infoscan and Behaviorscan it is possible to recover information on individual items and prices. For Behaviorscan the purchase data can be linked with the socioeconomic characteristics of the sample households. For Infoscan the price and quantity data can be linked to market characteristics such as type of clientele and other socioeconomic characteristics of the market area.

The data for Infoscan are available since 1985 and for Behaviorscan since 1978. Behaviorscan and Infoscan offer the researcher the opportunity to combine accurate purchase data from store scanners with socioeconomic information of purchasing households. Obtaining data from IRI for research purposes outside of IRI has proven to be difficult and costly.

PRAGMATIC ISSUES WITH ALTERNATIVE DATA SOURCES

Researchers attempting to use more than one of the above data series—even within the same major survey series—must exercise great care. The data are oftentimes not as comparable as they might at first appear. For example, in the 1972–74 Bureau of Labor Statistics Consumer Expenditure Survey, students were not included or surveyed if living away from home in a dormitory. In the 1980–81 sample they were included as separate consuming units. Other well-known examples include when or if food purchased and consumed away from home is reported; butter is included in fats and oils in the 1977–78 USDA Household Food Consumption Survey and in dairy products in the recent BLS surveys; sales taxes can be included or excluded from the cost of the product; the inclusion of one-person households varies across surveys and within surveys from one period to another. Probably the most troublesome variable is income. Both its definition and reference period depend upon the particular survey. In fact, the reference period for annual income varies from one sample household to the next in the most recent BLS surveys. In the 1977–78 NFCS, income data included the income of all household members ages 14 and older except roomers, boarders, and employees. In contrast, in the 1965 NFCS income data were recorded only for all related persons living in the household who were part of the family's finances.

Changes in the definition of standard statistical areas can produce codes of metro-nonmetro, farm-nonfarm, and urban-rural that do not include the same areas as in earlier surveys. Up through 1977–78 the NFCS distinguishes between central city (areas with populations of 50,000 or more in SMSAs), suburban (areas within boundaries of an SMSA but not in a central city), and nonmetropolitan (all areas not within an SMSA). The definition of the SMSA is based upon the 1970 definitions of areas by the U.S. Department of Commerce based on the 1970 population census. The CES codes a unit as inside or outside an SMSA and urban or rural. The CES defines urban as all persons living in SMSAs and in urbanized areas and urban places of 2,500 or more persons outside of SMSAs. Those not living in urban areas are classified as rural. New definitions of statistical areas were established in 1983. As a consequence many area titles were changed and geographic boundaries redefined. Previously, the standard metropolitan statistical area (SMSA) and standard consolidated statistical areas (SCSA) had been the categories. After 1983 there are three categories: metropolitan statistical area (MSA), primary metropolitan statistical area (PMSA), and consolidated metropolitan statistical area (CMSA). An MSA is an urban area that meets a specified size criteria, a PMSA is an urban area within a very large metropolitan area, and a CMSA is a combination of contiguous metropolitan areas.

The basic respondent unit of the CES is a consuming unit (CU). A CU consists of a family, or two or more people living together who pool their income to make joint expenditure decisions, or a person living alone, sharing a household with others, or living as roomer in a private home or lodging house, hotel or motel, but who is financially independent. To be financially independent a respondent must expend on two out of the three expense categories of housing, food, and other living expenses. In the CES there can be more than one CU in a household. In the NFCS a household consists of all the persons who occupy a house, an apartment, or other groups of rooms that constitute a housing unit. The sample does not include rooming houses, hotels, motels with permanent residents, or other group living establishments. Only housekeeping households are included in the final data set. A housekeeping household is one with at least one person having ten or more meals from the household's food supply during the seven days preceding the interview. The point is that a CU or household in the CES is not comparable to one in the NFCS and what is defined as a household or CU in one survey may have changed by the next survey.

In earlier surveys the CES used the term "head of household" and identified that person as the male in the household if both male and female were present. Since 1980 the term has been changed to "reference person." The reference person is the person who owns or rents the home. If the abode is jointly owned, the first person listed becomes the reference person. In the NFCS the male is used as the head of the household if both male and female are present.

The researcher should carefully examine changes in the definition of demographic characteristics such as race and origin, employment status, and occupation across data sets. Many other examples could be cited. The important point is that researchers must fully understand the definition of each variable *and* the sampling procedure before making comparisons.

Documentation for the data set is an absolute necessity but, by itself, is not sufficient if the data set is to be properly used. Complete documentation, particularly of survey data, is oftentimes lacking and may not correspond exactly with the data set in hand. Major consumer expenditure or consumption surveys usually provide documentation in the form of a codebook that lists the variables in the data file and assigned values. It is not unusual to find discrepancies between the published codes and those encountered in the data set. *It is recommended that the data codes be checked against marginal distributions of the variables.* Such descriptions can be important sources of information on added codes, changes in definitions, weighting procedures, level of nonresponse, and data checking and cleaning procedures. Different groups are responsible for preparing the documentation and the machine-readable data. The division of labor may create discrepancies between the documentation and the actual data. Variable distributions will reveal codes and values not included in the documentation and avoid later surprises in analysis of the data.

Finally, in spite of all the best efforts of the agency generating the data, there can still be problems with large or unusual data values that can seriously distort a statistical estimate. Very large expenditures are particularly troublesome. Should they be included or excluded? From the 1977–78 USDA/NFCS some examples are: $50,000 of welfare income, $35,000 of unemployment benefits, two-week purchases of 13.5 pounds of tea, 21 pounds of coffee, 18 pounds of lard, 30 pounds of baby food, 103 pounds of nondairy creamers and toppings, 300 pounds of fresh fruit purchased in the winter quarter, 500 pounds of pudding, or expenditures of more than 100 percent of weekly income on fresh whole milk. Economic models assume positive household income, yet more than 1,600 households in the 1977–78 NFCS reported negative annual incomes. For more examples see Buse (1979) and Buse and Johnson (1986).

Even scanner data, automatically recorded by sophisticated equipment, are subject to inaccuracies. Sales are not recorded and poorly trained checkers, bad labels, and incorrect UPC codes contaminate the data set (Lesser and Smith 1986). The point is that the researcher must clearly decide what to do with such observations and to assess the implications for the ensuing analysis.

> The data are imperfect not by design, but because that is all there is. Empirical economists have over generations adopted the attitude that having bad data is better than having no data at all, that their task is to learn as much as is possible about how the world works from the unquestionably lousy data at hand. While it is useful to alert users to their various imperfections and pitfalls, the available economic statistics are our main window on economic behavior. In spite of the scratches and the persistent fogging, we cannot stop peering through it and trying to understand what is happening to us and to our environment, nor should we. The problematic quality of economic data presents a continuing challenge to econometricians. It should not cause us to despair, but we should not forget it either (Griliches 1985, p. 199).

NOTES

1. The rural income maintenance experiment and the New Jersey graduated work incentive experiment conducted by the Institute for Research on Poverty at the University of Wisconsin used the experimental approach to measure the impact of income support programs on work incentives. They are two of very few real world examples of the experimental approach in economic analysis and modeling. The data set contains expenditures on housing, durables, clothing, health, and a 24-hour recall of food intake (Bawden 1970).
2. Expenditures are also used as the dependent variable. If equation (9.1) is multiplied by its

own price, then the right-hand side variable is expenditure. For details on alternative models see Chapters 4 and 7.

3. See Meyer and Kuh (1957) for further discussion. Kuznets (1966) says it is inappropriate to use cross-sectional data to make inferences about past long-term trends. For forecasting purposes time-series data is most appropriate.
4. The terms "household" and "consumer unit" are used interchangeably. However, the consumer unit is the most appropriate term. The exact definitions are in the *BLS Handbook of Methods*, Bulletin No. 2134 (1982) or BLS Bulletin No. 1992 (1978).
5. The data on individuals and households are only available in machine-readable format for the spring of 1965 (approximately 7,500 households out of a total sample of 15,000).
6. Home supplies include food and beverages used at home whether eaten at home, carried from home in packaged meals, thrown away, or fed to pets. Excluded food included commercial pet food, household food fed to animals raised for commercial purposes, food that was given away for use outside the home, and food consumed at restaurants, fast-food outlets, roadside stands, and meals at other homes.
7. Although the data can be compared with the 1977–78 NFCS there are differences in data collection procedures, food composition information, and nutrient data base used to calculate nutrient intake. For details see USDA (1985).

REFERENCES

Altonji, Joseph G. and Aloysius Siow. 1987. "Testing the Response of Consumption to Income Changes with (Noisy) Panel Data." *The Quarterly Journal of Economics* (May): 293–328.

Basiotis, P., M. G. Brown, S. R. Johnson, and K. J. Morgan. 1983. "Nutrient Availability: Food Costs, and Food Stamps." *American Journal of Agricultural Economics* 65:685–693.

Bawden, Lee. 1970. "Income Maintenance and the Rural Poor: An Experimental Approach." *American Journal of Agricultural Economics* 52:638–644.

Benus, J., J. Kementa, and H. Sapiro. 1976. "The Dynamics of Household Budget Allocation to Food Expenditure." *The Review of Economics and Statistics* 57(May):129–138.

Blanciforti, Laura A. and Ralph Parlett. 1987. "Changes in the CPI." *National Food Review* Spring:13–17.

Boehm, William T. and E. M. Babb. 1975a. *Household Consumption of Beverage Milk Products*. Purdue Agricultural Experiment Station, Bulletin No. 75, March.

_____. 1975b. *Household Consumption of Storable Manufactured Dairy Products*. Purdue Agricultural Experiment Station, Bulletin No. 85, March.

Bobst, Barry W., Robert E. Branson, Richard C. Haidacher, Eva E. Jacobs, Robert Raunikar, Benjamin J. Senauer, David Smallwood, Daniel S. Tilley, and Lilian R. de Zapata. 1987. "Data Sources for Demand Analysis." In *Problems, Issues, and Empirical Evidence on Food Demand and Consumption in the United States*, edited by Robert Raunikar and Chung-Liang Huang. Ames: Iowa State University Press.

Brown, Mark G. and Jong-Ying Lee. 1986. "Orange and Grapefruit Juice Demand Forecasts." In *Food Demand Analysis: Implications for Future Consumption*, edited by Oral Capps, Jr., and Benjamin Senauer. Blacksburg, Va.: Department of Agricultural Economics, Virginia Polytechnic Institute.

Buse, Rueben C. 1979. "Data Problems in the BLS/CES PU Diary Tape: The Wisconsin 1972–73 CES Diary Tape." Agricultural Economics Report #164. Madison: Department of Agricultural Economics, University of Wisconsin, July.

Buse, Rueben C., and A. C. Johnson, Jr. 1986. "Diagnosing a Data Base for Demand Analysis." In *Workshop on Demand Analysis and Policy Evaluation*, edited by D. P. Stonehouse. International Dairy Federation. Bulletin No. 197. Brussels, Belgium.

Buse, Rueben C., Thomas Cox, and John Glaze. 1985. "Structural Changes in the Demand for

Food." In *Consumer Demand and Welfare: Implications for Food and Agricultural Policy*. Edited by Jean Kinsey, North Central Regional Research Publication No. 311. University of Minnesota Agricultural Experiment Station. St. Paul, Minnesota.

Cagan, Phillip and Geoffrey H. Moore. 1981 *The Consumer Price Index: Issues and Alternatives*. Washington, D.C.: American Enterprise Institute for Public Policy Research.

Capps, O., Jr. 1987. "The Ultimate Data Source for Demand Analysis?" Paper presented at the 1987 AAEA Meetings, Michigan State University, East Lansing, Michigan, August.

Capps, O., Jr. and J. Havlicek, Jr. 1981. *Meat and Seafood Demand Patterns: A Comparison of the S_1-Branch Demand System and the Constant Elasticity of Demand System*. Bulletin No. 81-2. Blacksburg, Va.: Virginia Polytechnic Institute. Agricultural Experiment Station.

Capps, O., Jr., G. D. Spittle, and T. Finn. 1981. *The Virginia Tech Version of the 1972–74 BLS Consumer Expenditure Dairy Survey: Data Description and Data Inconsistencies*. Staff Paper SP-81-4. Blacksburg, Virginia Polytechnic Institute. Department of Agricultural Economics, April.

Carlson, Michael D. 1974. "The 1972–73 Consumer Expenditure Survey." *Monthly Labor Review* December, 16–23.

Coordinating Committee on Evaluation of Food Consumption Surveys. 1984. *National Survey Data on Food Consumption: Uses and Recommendations*. Washington, D.C.: National Academy Press.

Cox, Thomas L. and Michael K. Wohlgenant. 1986. "Prices and Quality Effects in Cross-Sectional Demand Analysis." *American Journal of Agricultural Economics* 68:908–919.

Duncan, Gregory and Daniel Hill. 1984. "An Investigation of the Extent and Consequences of Measurement Error in Labor Economic Survey Data." Survey Research Center, University of Michigan, July.

Folwell, R. J. and J. L. Baritelle. 1978. *The U.S. Wine Market*. Washington, D.C.: USDA/ESCS, AER-417, U.S. Government Printing Office.

Food and Agricultural Organization of the United Nations. 1981. *Bibliography of Food Consumption Surveys*. FAO Food and Nutrition Paper 18, Rome, Italy.

_____. 1986. *Review of Food Consumption Surveys—1985*. FAO Food and Nutrition Paper 35, Rome, Italy.

Forker, Olan D., Donald Liu, and Susan Hurst. 1987. "Dairy Sales Data and Other Data Needed to Measure Effectiveness of Dairy and Advertising." Report to the National Dairy Promotion and Research Board. Ithaca: Department of Agricultural Economics, Cornell University.

Funk, T. F., Karl D. Meilke, and H. B. Huff. 1977. "Effects of Retail Pricing and Advertising on Fresh Beef Sales." *American Journal of Agricultural Economics* 59:533–537.

Garkey, Janet and Wen S. Chern. 1986. *Handbook of Agricultural Statistical Data*. College Park, Md., Department of Textiles and Consumer Economics, University of Maryland.

Gieseman, Raymond W. 1986. "The Consumer Expenditure Survey: Quality Control by Comparison Analysis." *Monthly Labor Review* March, 8–14.

Griliches, Zvi. 1985. "Data and Econometricians—The Uneasy Alliance." *American Economic Review* 75:196–200.

Intriligator, Michael D. 1978. *Econometric Models, Techniques, and Applications*. Englewood Cliffs, NJ: Prentice-Hall.

Ireland, Lola M. 1972. "Retirement History Study: Introduction." *Social Security Bulletin* 35 (November):3–8.

Jourdan, Donna Kathryn. 1981. "Elasticity Estimates for Individual Retail Beef Cuts Using Electronic Scanner Data." M. S. Thesis, Texas A&M University.

Judge, George G., William E. Griffths, R. Carter Hill, and Tsoung-Chao Lee. 1980. *The Theory and Practice of Econometrics*. New York: Wiley.

Kokowski, Mary F. 1986. "An Empirical Analysis of Intertemporal and Demographic Variations in Consumer Preferences." *American Journal of Agricultural Economics*

68:894–907.

Kuznets, S. 1966. *Modern Economic Growth*. New Haven: Yale University Press.

Lesser, Willian G. and Jonathan Smith. 1986. "The Accuracy of Supermarket Scanning Data: An Initial Investigation." *Journal of Food Distribution Research* 17 (February):69–74.

Maddala, G. S. 1983. *Limited Dependent and Qualitative Variables in Econometrics*. Cambridge: Cambridge University Press.

Marion, B. W. and F. E. Walker. 1978. "Short-run Predictive Models for Retail Meat Sales." *American Journal of Agricultural Economics* 60:667–673.

Meyer, J. and E. Kuh. 1957. "How Extraneous Are Extraneous Estimates?" *Review of Economics and Statistics* 39:380–393.

Mundlak, Y. 1978. "On the Pooling of Time Series and Cross Section Data." *Econometrica* 46:69–85.

Nelson, Kenneth E. and Lawrence A. Duewer. 1986. "ERS's Measures of Red Meat Consumption". Paper presented at a Symposium on the Demand for Red Meat, Charleston, S.C., October 20–21.

Purcell, Joseph C. and Robert Raunikar. 1971. "Price Elasticities from Panel Data: Meat, Poultry and Fish." *American Journal of Agricultural Economics* 53:216–221.

Quackenbush, G. G., and James D. Shaffer. 1960. "Collecting Food Purchase Data by Consumer Panel." Bulletin 279. Michigan Agricultural Experiment Station Technical.

Raunikar, Robert and Chung-Liang Huang, eds. 1987. *Food Demand Analysis: Problems, Issues, and Empirical Evidence*. Ames: Iowa State University Press.

Rathje, William L. 1984. "Where's the Beef?" *American Behavioral Scientist* 28:(Sept./Oct.):71–91.

Social Security Administration. 1976. *1976 Almost 65: Baseline Data from the Retirement History Study*. Washington, D.C.: Office for Research and Statistics, Government Printing Office.

Swan, Patricia B. 1983. "Food Consumption by Individuals in the United States: Two Major Surveys." *Annual Review of Nutrition* 3:413–432.

U.S. Bureau of Labor Statistics. 1978. *Consumer Expenditure Survey: Integrated Diary and Interview Survey Data, 1972–73*. Bulletin 1992. Washington, D.C.: Government Printing Office.

_____. 1982. *BLS Handbook of Methods*. Bulletin 2134. Washington, D.C.: Government Printing Office.

_____. 1986. *Consumer Expenditure Survey: Interview Survey, 1982–83*. Bulletin 2246. Washington, D.C.: U.S. Government Printing Office.

U.S. Department of Agriculture. 1982. *Food Consumption: Households in the United States, Spring 1977*. Rpt. No. H-1. Washington, D.C.: U.S. Government Printing Office.

_____. 1985. *Nationwide Food Consumption Survey, Continuing Survey of Food Intakes by Individuals, Women 19–50 Years and Their Children 1–5 Years, 1 Day*. Report No. 85-1, Washington, D.C.: U.S. Government Printing Office.

_____. 1987. *Food Consumption, Prices, and Expenditures*. Statistical Bulletin No. 749. Washington D.C.: U.S. Government Printing Office.

_____. 1989. *Major Statistical Series of the U.S. Department of Agriculture: How They Are Constructed and Used. Ag. Handbook 671. Volume 5: Consumption and Utilization of Agricultural Products*. Washington, D.C.: Government Printing Office.

Wisniewski, K. 1984. "Statistical Issues in Using UPC Scanner Data". Proceedings of the American Statistical Association, Business and Economic Statistics Section. Washington, D.C.: American Statistical Association.

CHAPTER 10

Market Demand Structure for Dairy Products in the United States

Richard C. Haidacher

THIS CHAPTER focuses on the current status of the empirical market demand structure for dairy products in the United States. It begins with a brief discussion of the concept, role, and importance of demand structure and then provides a brief perspective on the essential components of empirical work directed toward estimating demand structure. Next, several selected sets of empirical estimates are presented for the traditional structural components of prices, income, and socioeconomic and demographic factors. Some of the salient characteristics of these empirical estimates are noted, and their economic significance is discussed. Finally, a brief assessment of existing empirical demand structure is provided.

ROLE AND IMPORTANCE OF DEMAND STRUCTURE

Few economic issues concerning national programs, policies, or forecasting can be adequately analyzed without some specification or assumption about the structure of consumer demand. For the dairy sector, the validity of this statement is evident throughout this volume. On the other hand, the relevant economic information for addressing these issues cannot often be obtained from knowledge of demand alone, since the demand structure is only one part of a larger, interdependent economic structure. The increased complexity of economic issues has focused

The views expressed are those of the author and do not necessarily represent those of the Economic Research Service or the U.S. Department of Agriculture.

attention on the importance of economic interdependence and forced a movement away from partial to more general equilibrium analyses. For example, in very partial equilibrium analysis a single own-price and/or income elasticity of demand might suffice, but a more general analysis may require a number of such estimates and the corresponding cross-price and income elasticities.

Recognition of these circumstances was clearly evident in the title of the now classic study by George Brandow (1961), *Interrelations among Demands for Farm Products and Implications for Control of Market Supply*. The content of this study left little doubt about the importance of these phenomena and it clearly indicated that requirements on both the amount and kind of information about demand structure had become more stringent. However, the advent of methods, such as those in Chapter 4, to provide this information have lagged and, as we shall see, the empirical estimates for dairy products have lagged even more.

Use of terms such as structure, demand structure, and market demand structure are widespread, appearing in both popular and technical literature. The meaning and connotation attributed to each of these terms varies considerably from one source to another. Therefore, it is useful to begin by asking what we mean by demand structure. Simply speaking, demand structure refers to the responses of consumers to various economic and related factors that ultimately produce the consumption behavior we observe; for example, the observed levels and changes in per capita consumption that result from changes in prices and income or are otherwise affected by factors such as age, season of the year, race, and such. Similarly, the domestic market demand structure for dairy products refers to the combined consumption response of all consumers in the United States to the various economic and related factors that determine the levels and changes in per capita consumption for the nation as a whole.

To visualize the complexity of such a demand structure, consider a table where all the factors influencing consumption are designated as column headings across the top and the various products are listed sequentially down the side to form rows, such that the body of the table is comprised of cells—one cell at the intersection of each row and column. Then if each cell represents a possible consumption response to a particular factor, the table of responses can represent the demand structure for specified values or ranges of the factors listed across the top. And, since the response can vary with the value or level of the factor (e.g., income), at least conceptually, the table should be extended in a third dimension to encompass the complete demand structure.

Thus, in this context, the complete set of consumption responses of all consumers in the United States to each of the economic and related factors comprises the "true" market demand structure. Alternatively, this conceptualization could be called the "theoretical" market demand structure. As such, it is only a theoretical abstraction; nevertheless, this or a similar characterization of demand structure is essential because it provides the conceptual framework within which most empirical research on demand structure is conducted. Therefore, as a basis for examining various empirical estimates, we briefly consider how the various components of empirical investigation relate to this framework.

BRIEF PERSPECTIVE OF EMPIRICAL WORK ON DEMAND STRUCTURE

To provide a perspective within which to examine the empirical estimates of demand structure we briefly consider the role and nature of the theory, data, objectives, and measurement methods employed in empirical analyses of demand structure. The economic theory of individual consumer demand addressed in Chapter 4 plays a strategic role, as described, for example, by Haidacher (1983). First, it is the primary foundation for almost all contemporary empirical analyses of consumer demand behavior and the basis from which the concept of true market demand structure originates. Second, it prescribes the majority of economic determinants comprising the market demand structure, and it provides certain specifications on the relevant data to be analyzed and the econometric techniques to be employed.

FACTORS AFFECTING CONSUMPTION

Economic theory of individual consumer demand specifies that the prices of all goods that comprise the budget of the individual consumer are potential factors affecting variation in quantities purchased of each individual item. Simply put, this phenomenon emanates from the consumer's budget constraint, or expenditure equation. This equation is simply the summation of each quantity purchased multiplied by its respective price, and the sum equals total expenditure on all goods and services. Consequently, for a fixed total expenditure, if any one price changes, the equality of the sum can only be maintained if one or more of the quantities are adjusted. Therefore, any or all of the quantities are potentially affected by a change in any one of the prices. Likewise, an increase in the level of total expenditure potentially affects all goods. Put another way, the quantities demanded are interdependent, each one depends potentially on each and every price and the level of expenditures.

Thus, the theory of individual consumer demand prescribes the set of all prices and individual income as the set of economic variables potentially affecting levels and changes in the quantity demanded of any good or commodity. And since there exists no specific theory of market demand, the aggregate demand of all individual consumers comprises the primary basis for specification of market demand and includes at least the same factors. In proceeding from the theory of the individual to market demand behavior, the distribution of total income among the population of consumers has been recognized as a relevant additional influence on market demand; but, more often than not, exigencies of the data preclude adequate consideration of this factor. It is in this context that the theory of individual consumer demand leads to the concept of true market demand structure and the specification of the complete set of economic determinants (Haidacher 1983).

In addition, a number of socioeconomic and demographic factors are often incorporated into demand analyses. Largely, the rationale is that these factors in some way influence tastes and preferences (or the preference function) underlying demand theory and thereby affect consumption behavior. Numerous empirical studies have provided substantial factual support for the influence of these variables on long-term consumption behavior, making analysis of their effects in demand research commonplace. Therefore, in addition to the traditional set of economic determinants cited above, the responses to this set of factors also comprise part of the true market demand structure. In addition to income-related factors, this set of variables includes population growth, age, sex, race, family size and composition, among others. The major variables from this set to be examined in this chapter are income, population changes, regional factors, age, race, and season of the year.

Other factors have been variously alleged to affect the domestic market demand for food commodities. These include various concerns related to health and nutrition, product convenience, food donations, and commodity advertising and promotion. To the extent that these factors can be adequately defined and conceptualized in terms of their influence on individual and/or market demand behavior, the responses to those factors also comprise part of the conceptual true demand structure. Likewise, the estimated responses would comprise part of the empirical demand structure.

However, relative to the research effort devoted to analysis of price, income, and demographic effects on market demand, most of these factors have been largely neglected, at least until relatively recent times. Consequently, in most cases adequate evidence is too sparse to support a comparative assessment with the empirical estimates of price, income, and demographic factors presented in this chapter to determine relative importance, magnitude, and so on. Also, the bulk of the recent work has focused on advertising and promotion of dairy products and, because of its special nature and current relevance, it is excluded here and treated in detail in other chapters.

OBJECTIVES OF ANALYSES

Ideally, one might conclude that the primary objective of empirical analyses of demand structure is that of obtaining the best possible estimates of consumer response parameters for application in analyses of programs and policy, in forecasting, and in evaluation of the relative economic significance of the structural elements. However, this ideal situation is precluded in empirical analyses for both practical and methodological reasons.

Methodologically, no one knows what the true demand structure is and there is no precise or definitive means of knowing if it has been found in any particular empirical instance. At best, we can obtain estimates of parts of that structure and, to a reasonable degree, we may be able to determine when we have not obtained it. Moreover, at any point in time or over any period of time, an ideal data set containing all the information required to construct the complete empirical structure of market demand does not exist. And, even if such a data set existed there is no unique empirical method or technique that would enable us to extract the comprehensive empirical structure. These circumstances exacerbate the practical difficulties of empirical research on demand structure.

Partly due to the above, and until relatively recently, most demand analyses focused on small parts of demand structure, and then, usually as a subsidiary objective to some policy problem or issue related to a particular agricultural commodity or sector. For example, with respect to the dairy sector, Rojko (1958) obtained estimates of retail dairy demand as one part of the econometric structure of the dairy sector; Heien (1977) estimated retail demands in a dairy sector model to analyze the cost of the U.S. price support program; and Novakovic and Thompson (1977) estimated retail demands in a U.S. dairy sector model to analyze the effect of manufactured milk imports. Thus, in practice most empirical studies of dairy demand have not focused on estimating the complete demand structure, but rather have focused on obtaining only those parts necessary to answer a particular program or policy question. The interdependence in demand between dairy and other food products received minor attention in such studies.

In other studies, where the focus was on one or more individual dairy products, even the interdependence between dairy products was neglected. However, beginning with the study by Brandow (1961) and subsequent studies by George and King (1971), Heien (1982), Huang (1985), and Heien and Wessells (1988), the issue of economic interdependence is addressed more fully.

The divergence between the objectives in the ideal situation and those inherent in the above studies should serve primarily as a basis for interpretation of the empirical results and not as a basis for criticizing particular studies. For most of the results reported in this chapter, the studies were conducted under the various practical and methodological constraints mentioned above in addition to those imposed by the available data and methods discussed below.

Types of Data

A prerequisite for obtaining the domestic demand structure is the availability of a data base compatible with this objective. Generally speaking, such a data base would contain observations over time from which one could obtain the per capita total amount of each individual product consumed, along with corresponding observations on the various factors determining the respective consumption levels such as prices, income, the various demographics, and so on. But no single data set exists that satisfies the criteria inherent in this objective. Therefore, various parts of the demand structure often have to be obtained from different, independent data sets. In fact, the necessary data to estimate some parts of the structure for individual commodities such as away-from-home consumption does not exist. Actually, there are two kinds of data available: They are commonly referred to generically as time-series and cross-sectional (for more detail see Chapter 8).

In general, publicly available time-series data consist of national estimates of annual consumption, derived as apparent per capita disappearance from annual production data. Although this estimated consumption is an imputed residual and not a direct observation of consumer behavior, it is the only available consistent measure of total commodity usage over time. Data on annual prices and average per capita income that are more or less compatible with this measure of consumption are available. However, corresponding data on socioeconomic, demographic, and other causal factors are either not available or severely deficient for analytical purposes. Consequently, time-series data are most commonly used to estimate price and income parameters for total per capita consumption of individual commodities.

Publicly available cross-sectional data are obtained from national surveys from a sample of domestic households. There are two major surveys that provide such data on food products. The Nationwide Food Consumption Survey (NFCS) is conducted about every ten years by the U.S. Department of Agriculture and, until 1980, the Consumer Expenditure Survey (CES) was also conducted at about ten-year intervals by the Bureau of Labor Statistics (BLS). In 1980, BLS initiated the Continuing Consumer Expenditure Survey (CCES) on an annual basis. The sampled households report all individual food expenditures and/or quantities purchased for consumption at home for a specified recent period, usually one or two weeks. Given the statistical nature of these samples, projected at-home expenditures/quantities of food products for the total population of households can be obtained. Unit price data corresponding to the expenditures/quantities are not obtained, but income data and detailed information on many socioeconomic and demographic characteristics of sampled households are obtained.

Thus, the cross-sectional data consist of (1) food quantities/expenditures consumed at home (excluding quantities consumed in restaurants, institutions, and other food consumed away from home) and (2) detailed information on socioeconomic and demographic characteristics of the households; but they do not include corresponding per unit price information. Consequently, even though these data cover only consumption at home, they are the primary source for estimating the socioeconomic and demographic aspects of demand structure.

Some private firms also produce cross-sectional data from panels of households they maintain such as Market Research Corporation of America (MRCA) and National Purchase Diary Research, Inc. (NPD). The data produced by such firms may contain additional information such as per unit prices in MRCA data; but generally, the product coverage is much less comprehensive than with the publicly available survey data. Nevertheless, because the data contain corresponding price, socioeconomic, and demographic information, they are increasingly used to estimate parts of the demand structure, particularly for dairy commodities consumed at home. Several estimates of demand responses presented later were obtained from MRCA data.

Analytical Methods

Among the group of U.S. food commodities, the demand for milk and dairy products has no doubt received its share of empirical economic research and analysis (Dash and Sommer 1984). To a large extent, this is a result of the fact that the dairy sector has long been a focal point of legislative policy and programs at both the state and federal level, which had the auxiliary effect of producing considerable publicly available data over a relatively long period of time. Another consequence of these circumstances is that empirical estimates of dairy demand structure are often obtained from studies that had primary objectives other than estimating the demand structure for dairy products, per se, although there are exceptions.

A wide range of approaches and econometric techniques have been employed, ranging from estimating single equation demand relations, to estimating multiequation models of the whole dairy sector or parts of it, to estimation of response parameters for dairy products within a complete demand system for all food commodities. Most early econometric work, prior to the fifties, used single equation demand models for analyzing both time-series and cross-sectional data on dairy products. This common approach was in large part due to limitations imposed by both the available computing technology and the stage of development in econometric theory and methods. Such single equation demand models, combined with a variety of econometric techniques, are still the most commonly used for analysis of cross-sectional data.

Given further advances in theory and technology, subsequent approaches emphasized multiequation structural econometric and simulation models of the dairy market sector, which were empirically estimated by a number of techniques ranging from ordinary least squares to various simultaneous-equation techniques. This approach/method is still rather widely used because of its versatility and flexibility in providing empirical representations of commodity sectors for both policy and forecasting purposes.

Of more recent vintage is the complete demand-system approach, implemented with a variety of estimation techniques (Brandow 1961; Huang and Haidacher 1983). The essence of this approach is to achieve a closer correspondence between the theoretical model of consumer demand and the empirical model of estimated behavior. This is accomplished by attempting to encompass the spectrum of commodities in the consumer's budget and incorporate the implied interdependence in the theoretical demand model. From a practical viewpoint a major objective is to empirically capture the effects of commodity interdependence, in particular, cross-commodity substitution. Additionally, the approach permits the inclusion of more factors, greater detail, and the ability to obtain more individual commodity parameter estimates such as the various cross-price elasticities.

Although the complete demand-system approach was first applied to demand for U.S. food commodities by Brandow in the early sixties, the empirical methods and techniques were not refined. It wasn't until the mideighties that the demand for individual dairy products received more attention, due to increased disaggregation in complete demand systems (Huang 1985; Heien 1982) and the application of complete demand systems to survey data (Kokoski 1986; Heien and Wessells 1988).

EMPIRICAL EVIDENCE

The body of empirical evidence accumulated over time on the U.S. demand structure for dairy products is rather large, covering various structural characteristics and a wide range of aggregation over products, length of response period, and consumer units. The most prevalent structural components estimated are the own-price and income elasticity of demand. Many of these estimates have recently been compiled by Kilmer (1989). Estimates of cross-price elasticities are few and estimates of socioeconomic and demographic effects for at-home food consumption are numerous, but in summaries such as Kilmer's (1989), neither of these two sets have been placed within the comprehensive concept of demand structure presented earlier. In presenting the evidence, we will try to maintain this perspective in addition to a similar perspective on the various aspects of aggregation. The empirical evidence on demand structure presented will primarily focus on (1) the national market demand for dairy products/commodities, (2) longer run rather than shorter run behavior, and (3) the more aggregative rather than less aggregative product categories.

Two major sets of structural estimates are presented. The first contains various estimates of price and income elasticities for different levels of commodity aggregation, different time periods, different components of per capita consumption (i.e., total versus at-home), and different data sources (time-series versus cross-sectional). The second contains estimates of the effects of selected socioeconomic and demographic factors obtained from household survey data on at-home consumption. For both sets, the initial focus is on estimates from research conducted in the Economic Research Service (ERS). This serves two purposes. It provides a basis for comparison with other estimates and it provides a more complete set of estimates where alternatives are unavailable. This approach does not imply that the ERS estimates are necessarily better than alternatives.

Price and Income Response Parameters

We first present the empirical estimates derived primarily from time-series data on total per capita disappearance of dairy products, citing their salient characteristics, estimation problems affecting signs and/or magnitude, data or conceptual difficulties, and the differences or similarities with alternative estimates that are more or less comparable. Subsequently, we present alternative estimates of both price and income elasticities obtained from survey data on at-home consumption. Public survey data have been largely used to obtain income response estimates (along with demographic effects) because these data do not contain explicit corresponding price observations. However, some studies have attempted to derive or otherwise obtain compatible price information to estimate price responses along with income and demographic effects. Still other studies have used private data sources containing corresponding price data to estimate both price and income responses. Therefore, we present the set of income responses obtained by ERS from public survey data and then examine and compare alternative price and income estimates from both public and private data sources.

Total Per Capita Consumption Estimates. In terms of the average U.S. consumer's total budget, expenditures on all food comprise less than 15 percent. As a group, dairy products account for about 3.5 percent of total expenditure and about 25 percent of the food budget. By comparison, red meat products account for a little less than 5.5 percent of the total budget and about 40 percent of the food budget. To provide a broad perspective on the relationship of dairy products to other items within the total budget, it is useful to start with estimates of the demand structure for dairy products as an aggregate within the context of a complete demand system that includes all food products and the nonfood sector. This comprehensive view provides a perspective on the economic relationship of dairy products to other food categories and nonfood expenditures that comprise the consumer's budget, and it further demonstrates the economic interdependence that exists among these food expenditure categories.

Table 10.1 contains estimates of price and income response parameters for the total market demand for dairy products obtained from two different complete demand systems that contain estimates of all price and income parameters for about a dozen composite food commodities along with a nonfood sector. The quantity data used to obtain these estimates are USDA time-series data on total per capita commodity disappearance, beginning in the late forties to early fifties, through the early eighties. The composite categories and the time periods covered vary only slightly between the two systems and otherwise the data are the same.

One system, called the direct demand system, estimates the quantity demanded as a function of all prices and income for each of the 12 food categories and one nonfood category. This representation follows closely the specification of demand systems in Chapter 4 and the concept of demand structure at the beginning of this chapter. Thus, the aggregate per capita consumption of dairy products is related to the price of dairy products, the prices of other food and nonfood items in the consumer budget, and income. This provides estimates of each direct, cross-price, and income elasticity.

The column of Table 10.1 labeled elasticity shows the estimated price elasticity between the quantity of dairy products demanded and each of the individual composite food commodity prices, nonfood price, and income. That is, −.3046 is the percentage change in the quantity of dairy products demanded for a 1 percent change in the price of dairy products, other factors such as prices and income remaining unchanged (ceteris paribus); .0198 is the percentage change in the quantity of dairy products demanded for a 1 percent change in the price of red meat products, ceteris paribus, and so on. The respective estimates of the standard error (in parentheses) provide a descriptive measure of how good the elasticity estimate is. The pragmatic rule employed here is that the estimated elasticity is better the more it exceeds the magnitude of its standard error.

A second system, called the inverse demand system, estimates the price of a commodity category as a function of all 14 composite food and nonfood quantities, and a scale variable. This system provides estimates of each direct and cross-price flexibility and a scale parameter. The practical rationale for estimating the demand structure in the inverse form rests on its use in certain policy and forecasting applications. The general question posed is one of determining the price change for a given commodity that results from a change in the quantity marketed of that commodity or some other commodity.

Table 10.1. Elasticity and flexibility estimates for all dairy products from composite food demand systems for the United States

Composite commodity Quantity/price	Elasticity		Flexibility	
(1) Dairy	−.3046	(.0303)	−0.609	(0.202)
(2) Meat	.0198	(.0167)	NA	NA
(3) Beef/Veal	NA	NA	0.084	(0.049)
(4) Pork	NA	NA	0.267	(0.051)
(5) Poultry	−.0392	(.0142)	−0.032	(0.040)
(6) Fish	.0273	(.0121)	0.028	(0.028)
(7) Eggs	.0208	(.0076)	−0.485	(0.107)
(8) Fats	−.0338	(.0118)	−0.259	(0.077)
(9) Fruits	.0152	(.0133)	0.055	(0.062)
(10) Vegetables	.0372	(.0109)	−0.071	(0.093)
(11) Processed Fruits/Vegetables	−.0278	(.0211)	0.025	(0.098)
(12) Cereal	.1656	(.0253)	0.080	(0.059)
(13) Sugar	−.0153	(.0120)	−0.228	(0.097)
(14) Beverages	.0191	(.0090)	0.052	(0.047)
(15) Nonfood	−.0675	(.0529)	1.092	(0.309)
(16) Income	.1832	(.0482)		
Scale parameter	NA	NA	−1.903	(0.332)
Constant	−.0118	(.0015)	−0.023	(0.006)
Budget share	.0349	NA	0.0357	NA

Source: Elasticities are from Huang and Haidacher (1983) and flexibilities are from Huang (1988).

Note: Elasticities are uncompensated. Flexibilities are compensated. Standard errors are in parentheses. NA = not applicable.

Thus, in the column labeled flexibility the direct price flexibility estimate of −0.609 indicates that, other quantities and income remaining unchanged (ceteris paribus), a 1 percent change in the aggregate per capita consumption of dairy products is associated with a change in the opposite direction of about .6 percent in the aggregate price of dairy products. In contrast, the estimated cross-price flexibility between the quantity of pork and the price of dairy products of 0.267 indicates that an increase of 1 percent in per capita consumption of pork, ceteris paribus, is associated with a change in the price of dairy products of about .267 percent in the same direction. Upon taking appropriate account of the sign, the remaining cross-price flexibilities can be similarly interpreted. An interpretation of the scale parameter can be found in Huang (1988).

The composite demand systems provide a broad perspective on the structure of demand for dairy products within the context of other food and nonfood products and clearly demonstrate the potential interdependence between both food and nonfood goods comprising the consumer's budget. Just as clearly, however, it does not provide detail on the economic interrelations among various individual dairy products or those products that may comprise important substitutes and compliments. Nor does it provide information on income and demographic effects. However, information on price effects for individual commodities can be derived from parameter estimates obtained in disaggregated empirical demand systems such as the one estimated by Huang (1985) containing 40 food commodities from which the estimates in Table 10.2 are extracted.

Table 10.2 extracts from the 40-commodity demand system the estimated direct, cross-price, and income elasticities for margarine and for the individual dairy product categories: fluid milk, evaporated and dry milk, frozen dairy products, cheese, and butter. The diagonal elements of the first six columns in Table 10.2 are the direct price elasticities (e.g., −.3319 for cheese). The last column contains the income elasticities and the remaining off-diagonal elements are the cross-price elasticities (e.g., .0743 is the cross-price elasticity between the quantity of fluid milk and the price of evaporated and dry milk; and .7125 is the cross-elasticity between the quantity of evaporated and dry milk and the price of fluid milk).

The number of cross-elasticities having large coefficients relative to their standard errors clearly demonstrates the economic interdependence among this group of products. But, other estimates indicate the opposite of what might be expected based on past empirical results. A case in point is the cross-elasticity between butter and margarine.

Some historical empirical results and a priori theoretical expectation would suggest that butter and margarine are close substitutes and probably were substitutes at one time. But the two cross-price elasticity estimates in Table 10.2 are both small in magnitude and smaller than their respective

Table 10.2. Direct, cross-price, and income elasticities for dairy products and margarine

Product	Price elasticity for						
	Fluid milk	Evaporated and dry milk	Frozen dairy products	Cheese	Butter	Margarine	Income elasticity
				Coefficient			
Fluid milk	−.2588	.0743	−.0904	.1026	.0020	.0169	−.2209
	(.1205)	(.0411)	(.0287)	(.0240)	(.0205)	(.0114)	(.0686)
Evaporated and dry milk	.7125	−.8255	.2742	−.1395	.0887	−.0510	−.2664
	(.3939)	(.2642)	(.1380)	(.1010)	(.1134)	(.0825)	(.2230)
Frozen dairy products	−.2530	.0785	−.1212	.0210	−.0453	.0284	.0111
	(.0792)	(.0397)	(.0848)	(.0369)	(.0432)	(.0186)	(.0580)
Cheese	.4531	−.0675	.0313	−.3319	−.2409	.0402	.5927
	(.1088)	(.0479)	(.0607)	(.1174)	(.0577)	(.0213)	(.1197)
Butter	.0138	.0803	−.1435	−.4609	−.1670	.0477	.0227
	(.1787)	(.1033)	(.1367)	(.1109)	(.1748)	(.0666)	(.1915)
Margarine	.2008	−.0656	.1250	.1097	.0665	−.2674	.1112
	(.1392)	(.1050)	(.0824)	(.0573)	(.0934)	(.1379)	(.1073)

Note: The upper numbers are uncompensated elasticity estimates. The numbers in parentheses are the standard errors.

Source: Huang (1985).

standard errors, suggesting that they are not close substitutes. One possible rationalization for this apparent paradox is that (1) commodities need not be close substitutes throughout the entire range of relative prices or income, and (2) given the trend in relative prices for the two commodities, most of the potential substitution had previously occurred so that during most of the sample period underlying the estimates in Table 10.2, quantity change due to substitution was minimal.

The estimates of Table 10.2 from the same complete demand system of 40 food commodities show the interdependent relationships among the traditional group of dairy-related commodities. But, significant relationships also occur between commodities in this traditional group and foods outside this group. Table 10.3 presents the complete set of estimated elasticities for the commodities of Table 10.2 from the complete demand system of 40 food commodities. The estimates from Table 10.2 are repeated to provide orientation and interpretation of Table 10.3.

Several salient features emerge from these estimates. As might be expected, a number of the estimated elasticities are smaller than their standard errors, indicating a lack of significant economic dependence. Conversely, a large number of the estimates clearly indicate strong economic interdependence. And, as with the margarine/butter example in

Table 10.2, some estimates do not conform to a priori theoretical expectations and/or conventional wisdom, and consequently, beg further explanation. Although there is no intent to address this issue, later remarks on the characteristics of the estimates may be relevant.

Other Time-series Estimates. Kilmer (1989) has recently compiled and reviewed a rather extensive set of price and income elasticity estimates for dairy products from published studies based on data from the early 1920s to 1986. Kilmer and others have commented on these empirical results (Thraen and Hahn 1989).

With respect to Table 10.1, comparable time-series estimates for the category of all dairy products were not available since most studies at least disaggregate fluid milk products from other dairy products. In fact, strict comparability is difficult to achieve even for the disaggregated categories in Tables 10.2 and 10.3, because alternative estimates differ with respect to commodity definition, time period covered, model specification, and estimation technique. As mentioned previously, in part this results from the fact that estimation of demand is often a subsidiary objective of any particular study. Nevertheless, selected alternative estimates of direct price and income elasticities are summarized in Table 10.4 for comparative purposes.

Characteristics of Annual Time-series Estimates. To demand analysts familiar with elasticity estimates for major agricultural commodities, perhaps the most striking characteristic, given the number of studies conducted, is the relative paucity of sustainable elasticity estimates that exists for the dairy sector. More specifically, a very small proportion of estimates is significant by the usual statistical criteria; for a fair number, appropriate evidence is not provided; and a number of (often cited) estimates are judgmental, that is, not derived by statistical means from price/quantity data. Although it would be premature to draw any firm conclusions based on the few studies that currently exist, it appears that the more recent studies employing complete demand systems, such as the one by Heien (1982) and those depicted in Tables 10.1 and 10.2, produce a greater proportion of price elasticity estimates that compare more favorably with the size of their respective standard errors. Of course, this is not to say that this approach is without its problems, some of which are common to traditional procedures. Perhaps the conjecture that the kind of economic interdependence captured by these estimates is too important a characteristic to be neglected in such analyses is appropriate.

Further support for this conjecture and substantive evidence on the validity of the estimates in Tables 10.1, 10.2, and 10.3 as approximations to the aggregate U.S. demand structure for dairy products is provided from

Table 10.3. Estimated price and income elasticities for dairy products

Price / Quantity	Cheese	Fluid milk	Other milk	Butter	Margarine	Frozen dairy products
(1) Beef and veal	−.2618 (.0939)	.0194 (.0270)	−.0117 (.1188)	−.0620 (.1102)	−.0163 (.0754)	−.0664 (.0424)
(2) Pork	−.0468 (.0603)	−.0242 (.0183)	−.1595 (.0725)	−.0420 (.0731)	.0956 (.0470)	−.0645 (.0249)
(3) Other meats	.4756 (.1045)	−.0375 (.0287)	−.1660 (.1819)	.0931 (.1517)	.2134 (.1394)	−.0451 (.0543)
(4) Chicken	−.0690 (.0704)	.0711 (.0202)	.1297 (.0835)	.3620 (.0896)	−.1150 (.0471)	.0767 (.0288)
(5) Turkey	.0727 (.0619)	−.0396 (.0153)	−.1018 (.0707)	−.0339 (.0804)	−.1140 (.0431)	−.0026 (.0270)
(6) Fresh and frozen fish	.0747 (.0680)	−.0244 (.0182)	.0729 (.0880)	−.0961 (.0897)	.0784 (.0591)	.0375 (.0294)
(7) Canned and cured fish	.0656 (.0722)	.0320 (.0228)	−.1418 (.1090)	−.2270 (.0986)	.1359 (.0721)	−.0136 (.0369)
(8) Eggs	.0613 (.0563)	−.0193 (.0161)	.0979 (.0632)	−.0387 (.0654)	.0098 (.0359)	.0504 (.0229)
(9) Cheese	−.3319 (.1174)	.1026 (.0240)	−.1395 (.1010)	−.4609 (.1109)	.1097 (.0573)	.0210 (.0369)
(10) Fluid milk	.4531 (.1088)	−.2588 (.1205)	.7125 (.3939)	.0138 (.1787)	.2008 (.1392)	−.2530 (.0792)
(11) Evaporated and dry milk	−.0675 (.0479)	.0743 (.0411)	−.8255 (.2642)	.0803 (.1033)	−.0656 (.1050)	.0785 (.0397)
(12) Wheat flour	−.1000 (.1292)	−.0565 (.0817)	−.0679 (.2976)	.0701 (.2155)	.0992 (.1579)	.0522 (.0923)
(13) Rice	.0080 (.0512)	.0387 (.0368)	.0001 (.1284)	−.1058 (.0827)	−.0643 (.0605)	−.0158 (.0357)
(14) Potatoes	−.0042 (.0303)	−.0230 (.0168)	.0349 (.0537)	−.0613 (.0357)	−.0009 (.0244)	.0021 (.0146)
(15) Butter	−.2409 (.0577)	.0020 (.0205)	.0887 (.1134)	−.1670 (.1748)	.0665 (.0934)	−.0453 (.0432)
(16) Margarine	.0402 (.0213)	.0169 (.0114)	−.0510 (.0825)	.0477 (.0666)	−.2674 (.1379)	.0284 (.0186)
(17) Other fats and oils	−.1489 (.0580)	.0020 (.0259)	.0543 (.1287)	−.1226 (.1190)	.1845 (.1714)	−.0245 (.0394)
(18) Apples	.0639 (.0365)	−.0242 (.0203)	.1599 (.0675)	−.0989 (.0534)	.1035 (.0439)	.0274 (.0280)
(19) Oranges	−.0157 (.0365)	.0100 (.0190)	.0483 (.0744)	−.1073 (.0538)	.1102 (.0503)	.0245 (.0278)
(20) Bananas	−.0052 (.0367)	−.0319 (.0169)	−.1898 (.0761)	.2149 (.0580)	−.0464 (.0509)	.0511 (.0285)
(21) Grapes	−.0403 (.0266)	−.0063 (.0115)	−.0241 (.0502)	−.0114 (.0417)	−.0641 (.0339)	.0186 (.0191)
(22) Grapefruit	−.0374 (.0194)	.0348 (.0109)	−.0982 (.0368)	−.0390 (.0306)	.0741 (.0232)	.0318 (.0162)
(23) Other fresh fruit	.0716 (.0936)	−.0316 (.0497)	−.4096 (.2115)	.5255 (.1496)	−.3696 (.1421)	−.1216 (.0791)
(24) Lettuce	−.0287 (.0333)	.0240 (.0133)	.1290 (.0471)	−.1129 (.0474)	−.0172 (.0295)	.0042 (.0221)
(25) Tomatoes	−.0804 (.0329)	.0530 (.0132)	.0450 (.0651)	−.1267 (.0576)	.1003 (.0447)	.0233 (.0302)
(26) Celery	.0247 (.0119)	.0104 (.0048)	−.0312 (.0287)	.0194 (.0211)	−.0030 (.0216)	.0159 (.0087)

Table 10.3. *Continued*

Price / Quantity	Cheese	Fluid milk	Other milk	Butter	Margarine	Frozen dairy products
(27) Onions	−.0510 (.0236)	−.0052 (.0088)	.0006 (.0311)	−.0597 (.0310)	−.0108 (.0199)	.0020 (.0151)
(28) Carrots	−.0250 (.0252)	−.0240 (.0118)	.0598 (.0510)	.0005 (.0461)	−.0469 (.0341)	.0003 (.0203)
(29) Cabbage	.0244 (.0137)	−.0053 (.0054)	.0280 (.0233)	−.0168 (.0196)	.0368 (.0156)	.0153 (.0099)
(30) Other fresh vegetables	−.0373 (.0599)	.0195 (.0257)	.0345 (.0972)	.2184 (.0885)	−.0682 (.0607)	.0064 (.0430)
(31) Fruit juice	−.0270 (.0262)	−.0111 (.0108)	−.0089 (.0266)	−.0332 (.0277)	−.0223 (.0168)	−.0264 (.0137)
(32) Canned tomatoes	.0644 (.0199)	−.0032 (.0088)	.0125 (.0470)	−.0158 (.0378)	.0095 (.0310)	−.0145 (.0186)
(33) Canned peas	.0176 (.0212)	.0073 (.0081)	−.0534 (.0529)	.0767 (.0356)	−.0348 (.0394)	−.0099 (.0159)
(34) Canned fruit cocktail	−.0149 (.0232)	−.0071 (.0089)	.0654 (.0573)	.0267 (.0441)	.0884 (.0447)	.0166 (.0170)
(35) Dried beans, peas, nuts	.0172 (.0207)	−.0136 (.0083)	−.0138 (.0223)	−.0015 (.0209)	−.0053 (.0135)	−.0206 (.0109)
(36) Other processed fruits and vegetables	−.0820 (.1023)	.0448 (.0418)	.0699 (.1237)	−.0298 (.1170)	−.0090 (.0804)	.1150 (.0550)
(37) Sugar	.0364 (.0229)	.0101 (.0090)	.0008 (.0271)	.0508 (.0285)	.0255 (.0153)	.0069 (.0168)
(38) Sweeteners	−.0976 (.0545)	−.0278 (.0212)	−.0065 (.0594)	−.0810 (.0719)	.0177 (.0344)	.0197 (.0383)
(39) Coffee and tea	.0246 (.0229)	.0196 (.0086)	.0090 (.0251)	.0322 (.0277)	−.0115 (.0140)	−.0179 (.0145)
(40) Frozen dairy products	.0313 (.0607)	−.0904 (.0287)	.2742 (.1380)	−.1435 (.1367)	.1250 (.0824)	−.1212 (.0848)
(41) Nonfood	−.4064 (.3957)	.3933 (.2081)	.3066 (.8070)	.4401 (.6073)	−.6436 (.4801)	.1257 (.2504)
Income elasticity	.5927 (.1197)	−.2209 (.0686)	−.2664 (.2230)	.0227 (.1915)	.1112 (.1073)	.0111 (.0580)

Source: Huang (1985).
Note: Elasticities are uncompensated. Standard errors are in parentheses.

Table 10.4. Alternative post–World War II estimates of own-price and income elasticities for dairy products

Source	Time period	Commodity	Elasticity[a] Price	Income
Rojko	1947–54	Fluid milk	−0.410	0.410
Wilson and Thompson	1947–63	"	−0.310	0.340
Brandow	1961[b]	"	−0.285[c]	0.160
Prato	1950–68	"	−0.105	
George and King	1971[b]	"	−0.350	0.380
Heien	1947–79	"	−0.539+	−0.550+
Huang	1953–83	"	−0.259+	−0.221+
Rojko	1947–54	Cheese	−0.900	−0.790
Brandow	1961[b]	"	−0.70[c]	0.450
George and King	1971[b]	"	−0.460	0.250
Huang	1953–83	"	−0.332+	0.593+
Brandow	1961[b]	Ice Cream	−0.55[c]	0.350
George and King	1971[b]	"	−0.530	0.330
Huang	1953–83	Frozen deserts	−0.121	−0.011
Brandow	1961[b]	Butter	−0.85[c]	0.330
George and King	1971[b]	"	−0.650	0.320
Heien	1947–79	"	−1.926	−2.290
Huang	1953–83	"	−0.167	0.023
Brandow	1961[b]	Evaporated and condensed	−0.300[c]	0.000[c]
George and King	1971[b]	Evaporated	−0.320	0.000
Huang	1953–83	Evaporated and dry milk	−0.826+	−0.266

Sources: Rojko (1958), Wilson and Thompson (1967), Brandow (1961), Prato (1973), George and King (1971), Heien (1982), and Huang (1985).

[a] + indicates the coefficient would be statistically significant at least 10 percent level.

[b] The year indicates the date the study was published, not the time period covered by data.

[c] These are judgmental estimates, that is, they are not statistically derived from observations on prices, quantities, or income.

several simulation analyses performed with these estimates. Both statistical and graphic results (Huang and Haidacher 1983; Huang 1985) demonstrated very close conformity between simulated and actual data measurements on changes in per capita consumption.

Perhaps with the exception of the own-price elasticities for fluid milk, another salient characteristic is the extreme variability of the elasticity estimates from alternative studies. This variability appears to be largely independent of economic model specification and/or statistical estimation technique. For the more comparable estimates, the closest agreement is among the estimates of own-price elasticity for fluid milk, which appear to be in the neighborhood of −.30.

Several dairy products exhibit long-term trends in per capita consumption and this appears to have influenced the empirical elasticity estimates, especially the income elasticities. Apparently, the estimation procedures have been unable to disentangle the effects of trend (either positive or negative) from the behavioral response to income. It should be noted that this comment is inapplicable to the income elasticities attributed to George and King, since their estimates were derived from survey data for at-home consumption only.

Given the difficulties with trend, there is clearly a question about the appropriate magnitude of the income elasticity, and in light of the more recent estimates of this parameter for fluid milk, perhaps there is also a question about its sign. For example, both Heien (1982) and Huang (1985) obtain income elasticities that are large in magnitude relative to their respective standard errors, and with negative signs. Similarly, cheese consumption is characterized by a long-term upward trend and Huang (1985) obtains an income elasticity of positive sign and large magnitude. Therefore, it appears prudent to consider other empirical evidence such as parameter estimates for at-home consumption from surveys before drawing conclusions.

Elasticity Estimates for At-home Consumption of Dairy Products. Because USDA disappearance data are imputed from production data and therefore not direct observations on consumer behavior, obtaining measurement of consumer response to other factors such as socioeconomic and demographic variables has proved to be infeasible, judging from the published research literature. Consequently, to obtain information and measurements on these components of demand structure, researchers have turned to alternative data sources both public and private. The primary source has been information obtained directly from households comprising a sample survey or panel.

The major public data sources are the Nationwide Food Consumption Survey (NFCS), the Consumer Expenditure Survey (CES), and, more recently, the Continuing Consumer Expenditure Survey (CCES). The basic information obtained in these surveys includes (1) quantities (NFCS) and/or expenditures (NFCS, CES, CCES) on a detailed list of food products purchased for consumption at home and (2) rather detailed information on socioeconomic, demographic, and income characteristics of each household. Unit price information is not provided, nor is any quantity/expenditure information by product category provided for food consumed away from home. Thus, for dairy product consumption, these survey data provide a basis for description, measurement, and insight regarding only the at-home part of consumer behavior. For this purpose, these data sources have provided a considerable amount of useful information about the demand structure for dairy products, including the effects of income, age, race, region, season, and other factors. The following exposition focuses on estimates of income responses obtained from these and other survey data along with some additional estimates of price response for at-home consumption of dairy products. A later section summarizes the influence of the demographic factors.

ERS Estimates. Table 10.5 provides income elasticity estimates for a broad range of dairy products obtained in four different studies conducted by the Economic Research Service using USDA and BLS household survey data. Specific designation of statistical significance for the elasticity estimates in Table 10.5 is not presented because the elasticities were often not estimated directly but derived from other directly estimated coefficients. With possible minor exceptions, however, the elasticities were derived from statistically significant parameter estimates. The results of these studies provide a basis for comparing differences and similarities of estimates obtained when model specifications, surveys, and time periods differ. The earliest study by Salathe (1979) uses 1972–73 CES while the latest study by Blaylock and Smallwood (1986) uses CCES data for 1980–81. The other two studies, Smallwood and Blaylock (1981) and Blaylock and Smallwood (1983), both use data from 1977–78. The 1983 study concentrated only on dairy products, whereas the other three studies covered a more comprehensive set of food categories with less detail on dairy products. For total dairy products, the income elasticities range from .07 by Blaylock and Smallwood (1983) to .16 by Salathe (1979). In addition to differences in survey data and time period, the two studies also differed in the inclusion of household characteristics as explanatory variables. The Salathe study included only income and household size, while the Blaylock and Smallwood study included a number of household characteristics. However, Smallwood and Blaylock (1981), in another study, which used the same model as Salathe but with NFCS data, found small elasticities similar to Salathe's. In yet another study, Blaylock and Smallwood (1986) used a different model (tobit) that included demographic variables and more recent CCES data than Salathe's (1979) but obtained elasticities almost identical to Salathe's. Thus, regardless of statistical technique, data, or model specification, changes in income appear to have limited effects on expenditures for total dairy products consumed at home.

For many individual dairy products, the income elasticities were also similar across the various studies. For example, it is noteworthy that fresh whole milk has a small negative expenditure elasticity (indicating that purchases decline as income increases). More expensive products, which typically are more processed, generally respond more to income changes than do less expensive products. Yogurt has a relatively large positive elasticity of .76, indicating that a 10 percent increase in income raises yogurt expenditures 7.6 percent.

Table 10.5. Income elasticities for at-home consumption from survey data

Dairy products	Study	Data	Income elasticities
Dairy products	Salathe (1979)	CES (1972–73)	.143[a]/.161[b]
	Blaylock and Smallwood (1986) (1980–81)	CCES	.138
	Smallwood and Blaylock (1981)	NFCS (1977–78)	.153
	Blaylock and Smallwood (1983)	NFCS (1977–78)	.072
Fresh milk products	Salathe	CES	.031[a]/.082[b]
	Blaylock and Smallwood (1986)	CCES	.021
	Blaylock and Smallwood (1983)	NFCS	−.009
Fresh whole milk	Salathe	CES	−.096[a]/−.043[b]
	Blaylock and Smallwood (1983)	NFCS	−.134
Other fresh milk products	Salathe	CES	.360[a]/.384[b]
	Blaylock and Smallwood (1983)	NFCS	.264
Processed dairy products	Salathe	CES	.312[a]/.274[b]
Processed milk	Smallwood and Blaylock (1981)	NFCS	−.084
	Blaylock and Smallwood (1983)	NFCS	−.016
Canned milk	Blaylock and Smallwood (1983)	NFCS	−0.118
Dry milk	Blaylock and Smallwood (1983)	NFCS	.075
Cream	Blaylock and Smallwood (1981)	NFCS	.527
	Blaylock and Smallwood (1983)	NFCS	.189
Butter	Salathe	CES	.290[a]/.179[b]
	Blaylock and Smallwood (1983)	NFCS	.241
	Blaylock and Smallwood (1986)	CCES	.350

Table 10.5. *Continued*

Dairy products	Study	Data	Income elasticities
Cheese	Salathe	CES	.387[a]/.370[b]
	Blaylock and Smallwood (1986)	CCES	.317
	Smallwood and Blaylock (1981)	NFCS	.321
	Blaylock and Smallwood (1983)	NFCS	.171
Natural American and Cheddar cheese	Blaylock and Smallwood (1983)	NFCS	.105
Processed cheese	Blaylock and Smallwood (1983)	NFCS	.126
Ice cream and related products	Salathe	CES	.317[a]/.302[b]
	Smallwood and Blaylock (1981)	NFCS	.241
	Blaylock and Smallwood (1983)	NFCS	.168
Yogurt	Salathe	CES	.759[a]/.605[b]
Dips	Smallwood and Blaylock (1981)	NFCS	.735
	Blaylock and Smallwood (1983)	NFCS	.451
Frozen and other dairy products	Blaylock and Smallwood (1986)	CCES	.211

Source: Haidacher et al. (1988).

Note: CES = Consumer Expenditure Survey, CCES = Continuing Consumer Expenditure Survey, NFCS = Nationwide Food Consumption Survey.

[a] Based on 1972 survey.

[b] Based on 1973 survey.

Alternative Estimates. Within the last two decades, researchers have increasingly used household panel data produced by private vendors to analyze various aspects of consumer behavior for dairy products. Boehm (1975), Thraen et al. (1978), and Blaylock and Blisard (1988) have used panel data from Market Research Cooperation of America (MRCA) to analyze the effect of various factors on the per capita consumption of selected dairy products. In part, the motivation for these efforts was to obtain measurements of demand responses to price, along with those of income and other factors, that were not readily extractable from public survey data that provided no direct price information. Alternatively, Kokoski (1986) and Heien and Wessells (1988) have attempted to use public survey data with derived price information in a demand systems framework to estimate price and income elasticities in conjunction with demographic effects. Table 10.6 summarizes some of the elasticity estimates from these studies. But, it should be remembered that these data also cover only the at-home part of consumption.

Characteristics of the Estimates. Overall, the most salient characteristics of the elasticity estimates for at-home consumption of dairy products is the rather large proportion of estimates that are statistically significant relative to those obtained for total per capita consumption from time-series data on disappearance. Of course, in part this is a statistical phenomenon that results from the substantially larger number of sample observations in survey data. One would expect this general result to hold for other commodities as well, and generally it does. But, as pointed out earlier for the time-series estimates, dairy products appear to have a relatively small share of robust elasticity estimates for own-price and income.

With respect to own-price elasticity estimates from survey data, most are statistically significant. But, within this subset the variation in magnitude is excessive, and this result occurs even when the basic data base and product definitions are the same. This implies that the results are quite sensitive to changes in model specification and/or specific data selection and editing and therefore not very robust or sustainable. Consequently, a good deal of caution should accompany any economic significance or confidence one attaches to these estimates.

In general, the income elasticity estimates in both Tables 10.5 and 10.6 appear consistent with the exception of those obtained by Kokoski (1986) and Heien and Wessells (1988) in a demand systems context. These latter estimates are based on food expenditures as opposed to total income or expenditure and therefore are not directly comparable. Although the remaining estimates vary somewhat from study to study, this can be largely attributed to factors such as different product definitions, model specifications, and such. Aside from this, there appears to be more consistency and convergence than inconsistency or divergence among the estimates. Especially noteworthy are the income elasticity estimates for fluid milk products. Most are relatively small in magnitude, some lack statistical significance and several estimates, using different data bases and time periods, find income elasticities with negative signs. This repeated occurrence of negative signs provides substantive support for the similar occurrence of negative sign obtained from total per capita disappearance data. However, given that two-thirds or more of fluid milk is consumed at

Table 10.6. Alternative elasticity estimates for at-home consumption of dairy and related products derived from cross-sectional data

Study	Data sources	Product	Elasticity[a] Own-price	Income
Salathe/Buse	NFCS 1965	All dairy products		0.146
Kokoski[b]	CES 1972–73	"	−0.973	0.797
"	CCES 1980–81	"	−1.013	0.819
Boehm	MRCA 1972–73	All fluid milk	−1.630*	0.050*
Thraen/Hammond/Buxton	MRCA 1972–73	"	−0.880*	0.120
Buse/Fleischner	NFCS 1977–78	"		0.048
Heien/Wessells	NFCS 1977–78	"	−0.630	0.770
Boehm	MRCA 1972–73	Whole milk	−1.700*	−0.070*
"	MRCA 1972–73	2% milk	−1.330*	0.160*
Buse/Fleischner	NFCS 1977–78	Cheese		0.320
Heien/Wessells	NFCS 1977–78	"	−0.520	1.010
Boehm	MRCA 1972–73	American cheese	−0.440*	0.160*
"	MRCA 1972–73	Processed cheese	−1.710*	0.120
Thraen/Hammond/Buxton	MRCA 1972–73	"	−0.620*	0.040
"	MRCA 1972–73	Natural cheese	−0.230*	0.250*
Boehm	MRCA 1972–73	Buttermilk	−1.520*	−0.170*
"	MRCA 1972–73	Ice cream	−0.420*	0.050*
"	MRCA 1972–73	Ice milk	−0.560*	−0.010
"	MRCA 1972–73	Yogurt	−0.510*	0.200
"	MRCA 1972–73	Cottage cheese	−1.290*	0.170*
Heien/Wessells	NFCS 1977–78	"	−1.100	1.020
Boehm	MRCA 1972–73	Butter	−0.760*	0.170*
Heien/Wessells	NFCS 1977–78	"	−0.730	1.060
Heien/Wessells	NFCS 1977–78	Margarine	−0.250	0.840
Boehm	MRCA 1972–73	Nonfat dry milk	−2.240*	−0.020*
Thraen/Hammond/Buxton	MRCA 1972–73	"	−1.920*	0.450*
Boehm	MRCA 1972–73	Canned milk	−1.330*	0.340*

Source: Except for Heien and Wessells (1988) the estimates were extracted from Kilmer (1989).

[a] Income elasticities for Boehm (1975) and Thraen et al. (1978) are quantity-income elasticities. Those for Salathe/Buse (1979) are expenditure-income elasticities. Those for Kokoski (1986) and Heien/Wessells (1988) are derived from estimated expenditure systems and can be characterized as quantity-expenditure elasticities where expenditure is the amount spent on food at home. An asterisk (*) indicates that the elasticity estimate would be statistically significant at the 10 percent level. For other estimates, direct evidence to make this determination was not available, even though the estimates may have been derived from statistically significant parameter estimates.

[b] These price and income elasticities are those computed and reported by Kilmer (1989). They are simple averages of selected estimates reported by Kokoski (1986).

home, the consistently small magnitude for the income elasticity from household data clearly indicates that the comparable estimates from disappearance data are too large.

Because the above results pertain only to the at-home consumption component of total consumption and because away-from-home consumption and use of dairy products as ingredients in other food products constitute important outlets for some dairy products, to maintain perspective it is useful to briefly sketch the nature of these characteristics for dairy products.

Drawing on data from both public and private sources, Haidacher et al. (1988) assessed the relationship between total dairy product disappearance and the components of at-home consumption, away-from-home consumption, and the use of dairy products as ingredients in other processed foods. About 37 percent of cheese is consumed at home, 39 percent away from home, and 24 percent as ingredients in food products. For butter, about 30 percent is consumed at home, 43 percent away from home, and 27 percent in other uses. For fluid milk products, about 68 percent is consumed at home, 16 percent away from home, and 15 percent as ingredients. Specific breakdowns for other dairy products were not reported.

Socioeconomic and Demographic Factors

Although the neoclassical theory of individual consumer demand does not explicitly prescribe socioeconomic and demographic variables as determinants of per capita food consumption, empirical evidence unequivocally demonstrates the influence of these factors on consumption. This is not to say that there is an absence of economic logic for inclusion of any of these factors in empirical demand analyses. On the contrary, it is well known that derivation of market demand from individual demand leads to market demand specifications related to the distribution of income. Nevertheless, household survey data have played an indispensable role in producing the accumulated evidence and in providing additional insights into consumer demand behavior for dairy products and other food commodities.

It is not feasible to inventory the volume of material that results from the combination of numerous empirical studies and over a half dozen socioeconomic and demographic variables. Instead, we will try to summarize some of the more salient features and implications of selected demographic variables. To accomplish this we focus primarily on the empirical results of the two ERS studies cited earlier by Blaylock and Smallwood (1983, 1986) and four demographic variables: age distribution, regional population distribution, race, and seasonal effects. Actually, with regard to the more prominent results, little generality is lost in focusing on a particular study because comparable studies differ only slightly on the magnitude of the effect of various household characteristics on demand and they are largely in agreement on the direction of effect. For example, the two studies by Blaylock and Smallwood (1983, 1986) using 1977–78 NFCS and 1980–81 CCES, respectively, found that per capita expenditures for dairy products were highest for residents of the Northeast and second highest for residents of the West; and that blacks spend considerably less for dairy products than do whites. Analyses by Salathe and Buse (1979) using 1965 HFCS data and by Salathe et al. (1979) using 1973 CES data obtained similar results. The results from the Blaylock and Smallwood study (1983) using 1977–78 NFCS data, which contain the most detail on individual dairy products, are presented in Table 10.7.

Age. Per capita dairy expenditures varied significantly across households with different-aged members. Households with children aged two and under spent more on total dairy products than did households without young children, mostly because of higher expenditures for fresh and processed milk (which includes infant formula). However, households with teenagers spent considerably more per person on total dairy products than did households with infants. Households with teenagers spent considerably more for fresh milk, frozen desserts, processed cheese, and sour cream and dips. Households composed of elderly persons had the highest per capita expenditures on other milk, processed milk, cream, and margarine. The elderly, however, spent less on cheese, especially processed cheese.

Race. Blacks and other racial groups (nonwhite, nonblack) had lower weekly per capita expenditures than did whites for virtually all dairy items. Blacks spent almost 25 percent less per person than did whites for total dairy products, 25 percent less for cheese, and 20 percent less for margarine. However, blacks spent approximately 86 percent more for canned milk. These findings are also consistent with those reported by Thraen et al. (1978).

Regional Distribution of the Population. Using the Northeast as a base, Table 10.7 shows how expenditures on dairy items vary across regions. Per capita expenditures on total dairy and related products are highest in the Northeast and lowest in the South. While this expenditure pattern was also true for cheese, it did not hold for all products. For example, other fresh milk expenditures were higher in the North Central, South, and West than in the Northeast. Natural American and Cheddar cheese expenditures were lowest in the Northeast, but northeasterners spent more for processed and other cheeses. Northeastern residents also spent more for butter and less on margarine than did residents of other regions. There appears to be slightly more similarity in dairy expenditures in the North Central, South, and West than across the nation. Differences in relative prices or dairy product standards may have caused some of the apparent differences in regional expenditures. For example, the South had higher cottage cheese prices, which may account for some of their lower expenditures on cottage cheese.

Table 10.7. Simulated differential effects of age, race, region, and season on weekly per capita dairy expenditures for at-home consumption

	Age[a]					Race[a]		Region[a]			Season[a]		
	0–2	3–12	13–19	20–39	65 and older	Black	Other Nonwhite/Nonblacks	North Central	South	West	Summer	Fall	Winter
							Percentage difference[b]						
Total dairy and related products[c]	15.2	−5.8	17.1	−2.6	0.1	−24.8	−6.8	−11.2	−14.0	−4.5	7.2	7.3	−6.5
Dairy products[d]	19.2	−3.6	20.6	.4	−1.8	−24.8	−6.4	−13.1	−15.4	−5.1	6.7	6.6	5.8
Fresh milk	35.7	20.5	46.5	−3.7	.3	−28.7	.7	−5.7	.6	−.4	5.3	8.2	9.6
Whole	20.5	22.8	45.6	−4.9	−5.2	−12.4	39.1	−37.6	−6.1	−32.9	7.5	13.9	14.2
Other	39.8	−21.6	10.4	−5.5	18.7	−48.5	−79.7	128.7	40.1	117.0	.3	−3.0	−.8
Processed milk	614.4	−31.8	11.5	−20.7	19.6	8.4	−39.1	−11.0	16.9	34.5	8.6	17.7	13.9
Canned	1,522.2	−49.0	−6.3	−43.9	37.2	86.5	3.7	9.3	121.5	91.8	6.2	19.4	1.3
Dry	45.2	−8.7	35.4	−2.1	11.3	−56.0	−72.0	−20.6	−33.4	16.0	20.5	19.2	35.2
Cream	−37.1	−19.1	−22.2	−38.7	62.3	−35.4	−57.9	20.5	−48.8	14.2	7.4	17.5	−2.0
Cream substitutes	−30.7	−20.6	−25.7	−57.2	56.6	−64.3	−57.9	14.8	24.4	−17.0	26.7	−8.4	2.0
Frozen desserts	−36.4	16.7	23.1	−19.5	14.6	−12.7	−22.3	−11.8	−16.4	−15.7	9.6	−11.4	−17.4
Cheese	−28.9	−8.5	11.2	22.9	−14.1	−25.0	−9.0	−15.1	−24.3	−1.5	5.4	8.8	8.4
Natural American and Cheddar	−19.8	−18.5	19.7	6.3	−7.7	−11.2	−20.5	26.0	24.4	98.4	3.1	8.5	8.5
Processed	16.4	30.1	23.5	8.1	−26.5	−27.1	−6.6	−8.8	−10.7	−34.5	8.6	9.4	19.3
Other	−45.4	−2.6	30.1	65.0	2.6	−63.0	−20.9	−45.5	−73.0	−34.1	4.9	5.1	−1.3
Cottage cheese	−76.2	−30.2	−19.7	−21.8	2.2	−67.8	−20.1	−4.9	−52.5	6.7	3.2	−9.8	−7.1

Table 10.7. *Continued*

	Age[a]					Race[a]		Region[a]			Season[a]		
	0–2	3–12	13–19	20–39	65 and older	Black	Other Nonwhite/Nonblacks	North Central	South	West	Summer	Fall	Winter
							Percentage difference[b]						
Sour cream and dips	−33.5	39.7	80.4	35.4	−7.5	−78.8	−45.5	−3.0	−43.8	7.4	1.9	4.6	9.4
Table spreads	−39.8	−34.4	−21.3	−13.3	10.5	−11.5	−11.1	−10.9	−22.7	−16.6	12.9	19.8	17.1
Butter	−54.8	−44.7	−19.2	22.2	−2.5	12.1	−16.3	−49.6	−69.8	−50.9	3.6	15.4	12.1
Margarine	−24.6	−23.9	−19.0	−31.1	17.5	−20.2	−10.2	22.5	16.5	13.6	14.7	18.5	16.6

Source: Blaylock and Smallwood (1983).

[a] Factors other than the specified variable are held constant at their sample means.

[b] Percentage change from expenditures in the base category. The base categories are: age: 40–64; race: Caucasian; region: Northeast; season: spring.

[c] Includes margarine and cream substitutes.

[d] Excludes margarine and cream substitutes.

Seasonal Variation. Weekly per capita dairy expenditures vary widely by season. Such differences may be caused by relative price variations, climate, and the timing of holidays. Expenditures were higher in the summer, fall, and winter than in the spring for total dairy and for many individual dairy products. For example, processed milk expenditures are highest in the fall and winter, probably because of holiday food preparation and baking. Per capita expenditures for frozen desserts were lowest in the fall and winter.

SUMMARY OF DEMOGRAPHIC EFFECTS

Socioeconomic and demographic factors such as regional, racial, and age distributions of the population, and seasonal purchase patterns influence dairy consumption. But even when combined, the region, race, and age factors have only a limited effect on yearly changes in per capita consumption (Blaylock and Smallwood 1986; Buse 1986; Salathe 1980; Musgrove 1982). Demographic variables are probably more important in explaining variations in expenditures between households or groups of households than in explaining yearly fluctuations in national per person spending because factors such as regional, racial, and age distributions change slowly over time (Blaylock and Smallwood 1986; Guseman and Sapp 1984; Buse 1986). Based on projections made with statistical results obtained from the 1980–81 CCES data (Blaylock and Smallwood 1986), Table 10.8 shows that even in the long run, changes in U.S. age distribution, regional population distribution, and racial composition would combine to increase per capita cheese consumption by less than 1.4 percent from 1980 to 2000 (column 3, last line). While population growth helps increase national dairy consumption, it cannot be relied upon to expand dairy consumption at historic rates because the population growth rate has slowed to less than 1 percent per year.

When combined with projected population growth, changes in demographic factors increase aggregate at-home demand for total dairy products by about 1 percent per year (Blaylock and Smallwood 1986). Cheese consumption at home is projected to increase about 11 percent during 1980–90, and milk and cream to increase about 10 percent. At-home consumption of total dairy products is projected to increase about 10.5 percent between 1980 and 1990 due to the combined effects of demographic changes and population growth.

RELATIVE IMPORTANCE OF PRICE, INCOME, AND DEMOGRAPHIC FACTORS

For the structural demand components including the set of relevant

Table 10.8. Projected changes in expenditures for food consumed at home due to shifts in demographics

Shifts	Milk and cream	Other dairy products[a]	Cheese	Butter	Total dairy products
	Changes in expenditures from 1980 levels				
			Percentage		
Age distribution					
1990	0.2	0.5	1.0	1.1	0.5
2000	−0.9	2.1	1.9	1.8	1.6
Regional distribution					
1990	−0.1	0.1	−0.2	−0.7	−0.1
2000	−0.2	0.3	−0.4	−1.3	−0.1
Racial distribution					
1990	−0.3	−0.2	−0.5	−0.2	−0.3
2000	−0.5	−0.4	−0.1	−0.3	−0.6
Total change[b]					
1990	−0.6	0.4	0.3	0.2	0.1
2000	−1.6	2.0	1.4	0.2	0.9

Source: Haidacher et al. (1988).
[a] Includes evaporated, condensed, and dry milk, and other dairy products.
[b] Net adjustment after accounting for projected changes in all demographic variables.

prices, income, and salient demographics, the empirical estimates presented provide evidence of the relative importance of the various structural components that affect per capita consumption. The estimated responses show the marginal effects of the various individual factors. For example, being independent of the units of measurement, the elasticities can be compared with respect to relative magnitude. Combining this information with other data and information on actual movements in prices, income, and demographic factors provides the basis for substantive conclusions on the economic significance of these factors. The general summary conclusion is that year-to-year changes in consumption are primarily the result of change in relative prices followed by changes in income and the demographic factors.

The relative impact of socioeconomic and demographic factors is evident in Table 10.8 and the associated discussion, supported by the cited studies. For income, the consistently small elasticity estimates for major product categories from household survey data imply that the magnitude of the corresponding time-series estimates are probably too large for fluid milk and cheese, where substantial long-term trends were inherent. Taking appropriate account of this exigency brings the results into conformity with the composite demand-system income elasticity estimate for all dairy products of about .18. The corresponding price elasticity estimate for all dairy products of about −.30 is substantially larger in magnitude. After accounting for the inflation of the income elasticity estimates due to trends for individual product categories, comparable relative magnitudes for own-price relative to income are evident. But, this accounting ignores the fact that prices other than own-price were found to have economically significant effects. To assess the full impact of the structural price and income components requires combining the estimated structure with actual (or projected) changes in the structural variables similar to the analysis underlying Table 10.8 for demographic factors. Simulation analyses conducted by Huang and Haidacher (1983), Huang (1985), Haidacher et al. (1988), Huang and Haidacher (1989), and Putler (1989) provide substantial statistical, graphical, and numerical evidence supporting the conclusion that own-price, other prices, and income are the predominant factors influencing per capita consumption.

CONCLUSION

The dairy sector of U.S. agriculture has received its share of economic research and analysis. Because demand structure is an essential part of most economic analyses, it has also received a good deal of attention. A large volume of empirical estimates have been produced in the process, although most of these estimates were not the result of a direct focus on estimating the demand structure for dairy products. Rather, estimation of the structural components of demand occurred as a subsidiary objective in most analyses. The remaining questions are what can we say about the status of knowledge on the structure of demand for dairy products and how can we improve it?

WHAT DO WE KNOW ABOUT DEMAND STRUCTURE?

Regarding total per capita consumption of dairy products, the estimates of own-price and income response for all dairy products appear to be reasonably consistent and perhaps about as good as the available data can provide given the level of aggregation. Too few estimates of the complete price structure exist for this level of aggregation to make an assessment, although simulation evidence provides some support. Upon disaggregation of total per capita consumption to product categories, the estimated elasticities vary considerably in their robustness and sustainability with concurrent variations in data and model specification and product definition. The most consistent elasticity estimates appear to be for the own-price of fluid milk. Own-price elasticities for manufactured products vary considerably in magnitude, precision, and sometimes even in sign (Haidacher 1986). Too few cross-price elasticities have been estimated and evaluated to provide an assessment, but the evidence clearly suggests that cross-commodity interdependence is economically significant and should not be ignored. As we have seen, income elasticities for total per capita consumption of individual dairy commodities can be significantly affected by trends inherent in the time-series data. Individual commodity estimates need to be closely scrutinized in conjunction with estimates from alternative data sources and then used with caution.

In contrast to the total per capita structural estimates, the estimates of income response for at-home consumption of dairy products appear to be superior to those for own-price response, showing consistency across different studies. Estimates of demographic responses appear to have a consistency similar to those for income. The reasons for the unsatisfactory price response estimates are not clear. In contrast, some of the reasons for the superior income and demographic response estimates reside with the applied expertise in these studies that has been developed over time. From an intuitive viewpoint at least, one is inclined to attribute a greater degree of credibility to these income and demographic response estimates because they are obtained from direct observation of consumer behavior. It is unfortunate that the responses only apply to at-home consumption behavior, which increasingly accounts for a smaller share for many products.

Major Discrepancies and Suggestions for Improvement

One could spend a good deal of time and effort compiling a list of deficiencies, the elimination of which would improve the analysis of demand and the estimation of demand structure. One could also spend a lot of time determining the specific priority ranking of the items, but the end result relies heavily on subjective criteria. In my judgment, for the dairy sector, the two major deficiencies concern data and methods.

The major data deficiency concerns the away-from-home consumption of dairy and other food products. Estimates of total food expenditure exist, but data to support demand analyses designed to estimate demand structure by product or commodity categories on a national basis are largely nonexistent. Even aside from the cost, removal or amelioration of this problem is not a straightforward issue. It is one that requires a good deal of prior study from a perspective that goes beyond the dairy sector. The methods discrepancy is not entirely one of techniques and is closely related to data issues. One concern focuses on the compatibility of disappearance data with the models of consumer demand used to estimate demand structure. The major question is whether the disappearance quantities (and associated prices) used in demand analyses correspond to the quantity and prices of the theoretical consumer demand model or whether the quantities more nearly correspond to the counterparts in the demand model for an input in a production process. This hypothesis appears to be relevant for dairy products that entail greater processing such as cheese (Haidacher 1986). Further study and resolution of this issue could pay large dividends in terms of improved estimates of demand structure.

Another discrepancy in the analysis and estimation of demand structure that has been largely ignored until recently is the economic interdependence among commodities and products, including those within the dairy sector. The existing evidence clearly demonstrates the economic significance of this phenomenon. The implication is that better estimates of demand structure will be contingent on incorporating this phenomenon into the analyses.

REFERENCES

Blaylock, James R. and David M. Smallwood. 1983. "Effects of Household Socioeconomic Factors on Dairy Purchases." USDA/ERS Technical Bulletin No. 1686.

_____. 1986. "U.S. Demand for Food: Household Expenditures, Demographics, and Projections." USDA/ERS Technical Bulletin No. 1713.

Blaylock, James R. and William N. Blisard. 1988. "Effects of Advertising on the Demand for Cheese." USDA/ERS Technical Bulletin No. 1752.

Boehm, William T. 1975. "The Household Demand for Major Dairy Products in the Southern Region." *Southern Journal of Agricultural Economics* 7(2):187–196.

Brandow, G. E. 1961. *Interrelations among Demands for Farm Products and Implications for Control of Market Supply*. Bulletin No. 680. Pennsylvania State University, College of Agriculture, University Park.

Buse, R. C. and A. Fleischner. 1982. "Factors Influencing Food Choices and Expenditures." Economics Issues No. 68, Department of Agricultural Economics, University of Wisconsin–Madison.

Buse, R. C. 1986. "Is the Structure of the Demand for Food Changing?" In *Food Demand Analysis: Implications for Future Consumption*, edited by O. Capps and B. Senauer. Blacksburg: Virginia Polytechnic Institute and State University.

Dash, Suzanne L. and Judith Sommer. 1984. *Recent Dairy Policy Publications with Selected Annotations*. U.S. Department Agriculture, Economic Research Service.

George, P. S. and G. A. King. 1971. *Consumer Demand for Food Commodities in the United States with Projections for 1980*, Gianinni Foundation Monograph No. 26. University of California–Berkeley.

Guseman, P. K. and S. G. Sapp. 1984. "Demographic Trends and Consumer Demand for Agricultural Products." *Southern Rural Sociology* 1:1–24.

Haidacher, Richard C. 1983. "Assessing Structural Change in the Demand for Food Commodities." *Southern Journal of Agricultural Economics* 15(1):31–37.

_____. 1986. "Developing Useful Models." In *Workshop on Demand Analysis and Policy Evaluation*, Bulletin No. 197 edited by D. Peter Stonehouse. Brussels, Belgium: International Dairy Federation.

Haidacher, Richard C., James R. Blaylock, and Lester H. Myers. 1988. *Consumer Demand for Dairy Products*. Agricultural Economics Report No. 586. U.S. Department Agriculture, Economic Research Service.

Heien, Dale M. 1977. "The Cost of the U.S. Dairy Price Support Program: 1949–74." *Review of Economics and Statistics* 59:1.

_____. 1982. "The Structure of Food Demand: Interrelatedness and Duality." *American Journal of Agricultural Economics* 64:213–221.

Heien, Dale M. and Cathy Roheim Wessells. 1988. "Dairy Products Demand: Structure, Prediction, and Decomposition." *American Journal of Agricultural Economics* 70(2): 219–28.

Huang, K. S. 1985. *U.S. Demand for Food: A Complete System of Price and Income Effects.* USDA/ERS Technical Bulletin No. 1714.

_____. 1988. "An Inverse Demand System for U.S. Composite Foods." *American Journal of Agricultural Economics* 70(4):902–909.

Huang, D. S. and R. C. Haidacher. 1983. "Estimation of Composite Food Demand System for the United States." *Journal of Economic and Business Statistics* 1(4):285–291.

_____. 1989. "An Assessment of Price and Income Effects on Changes in Dairy Consumption." In *Advertising, Promotion, and Consumer Use of Dairy Products,* edited by Cameron S. Thraen and David E. Hahn. Ohio State University.

Kilmer, Richard L. 1989. "Price and Income Elasticities for Dairy Products: A Research Overview." In *Advertising, Promotion, and Consumer Use of Dairy Products,* edited by Cameron S. Thraen and David E. Hahn. Ohio State University.

Kokoski, Mary F. 1986. "An Empirical Analysis of Intertemporal and Demographic Variations in Consumer Preferences." *American Journal of Agricultural Economics* 68(4):894–907.

Musgrove, Philip. 1982. *United States Household Consumption, Income, and Demographic Changes, 1975–2025.* Baltimore: Johns Hopkins University Press.

Novakovic, Andrew M. and Robert L. Thompson. 1977. "The Impact of Manufactured Milk Products on the U.S. Dairy Industry." *American Journal of Agricultural Economics* 59(3):507–519.

Prato, A. 1973. "Milk Demand, Supply, and Price Relationships, 1950–1968." *American Journal of Agricultural Economics* 55(2):217–222.

Putler, Daniel S. 1989. "An Overview of Dairy Product Consumption: An Executive Summary." In *USDA Report to Congress on the Dairy Promotion Program.*

Rojko, Anthony S. 1958. *The Demand and Price Structure for Dairy Products.* USDA Technical Bulletin No. 1168.

Salathe, Larry E. 1979. "Household Expenditure Patterns in the United States." USDA/ERS Technical Bulletin No. 1603.

_____. 1980. "Demographics and Food Consumption." *National Food Review* No. 10. Economics, Statistics, and Cooperative Service, U.S. Department of Agriculture.

Salathe, Larry E. and Rueben C. Buse. 1979. "Household Food Consumption Patterns in the United States." USDA/ERS Technical Bulletin No. 1587.

Salathe, Larry E., Anthony E. Gallo, and William T. Boehm. 1979. *The Impact of Race on Consumer Food Purchases* ESCS-68. Economics, Statistics, and Cooperative Service, U.S. Department of Agriculture.

Smallwood, D. and J. Blaylock. 1981. "Impact of Household Size and Income on Food Spending Patterns." USDA/ERS Technical Bulletin No. 1650.

Thraen, Cameron S. and David E. Hahn, editors. 1989. *Advertising, Promotion, and Consumer Use of Dairy Products.* Ohio State University.

Thraen, C. S., J. Hammond, and B. Buxton. 1978. "An Analysis of Household Consumption of Dairy Products." Station Bulletin No. 515. University of Minnesota Agr. Exp.

Wilson, R. R. and Russell G. Thompson. 1967. "Demand, Supply and Price Relationships for the Dairy Sector, Post–World War II Period." *American Journal of Agricultural Economics* 49:360–371.

CHAPTER 11

Bayesian Estimation and Demand Forecasting: An Application for the Netherlands

A. J. Oskam

THERE IS A wide variety of prediction or forecasting methods (Makridakis et al. 1983). Prediction methods with a scientific basis try to identify and quantify the factors that determine historic and future developments. The central idea behind such methods is that understanding the causal mechanism can improve the quality of predictions. Moreover, one can check afterwards which factors were responsible for differences between predictions and realizations. This may improve methods and results for future predictions. The prediction method we have in mind can be classified as the econometric or causal method (Makridakis et al. 1983).

Forecasting time-related developments can also be organized along different lines. Time-series methods using smoothing, decomposition, or autoregressive moving average (ARMA) are quite different from the econometric method. Here forecasts are based only on historic observations of the particular variable or on observations of a limited number of related variables.

Chapter 5 of this book outlines some alternative methods of forecasting demand. Here we attend to a specific method for forecasting demand, a method that uses information from several sources. It is related to the econometric method; however, information from other sources can be combined with sample information, using a Bayesian approach. We describe and develop an applicable method for such an approach. Different ways to incorporate prior information are discussed.

We will start with the specification of a demand function. A sufficient theoretical basis for demand functions together with enough flexibility in functional form to represent changes in preferences of consumers is an important starting point for an empirical analysis. Because demand theory has been discussed in Chapter 4, only limited attention will be given to the specification of a demand function.

After the introduction and discussion of a Bayesian method, it will be applied to the prediction of demand for dairy products in the Netherlands. This will be accomplished through reviewing the performance of different methods for estimating future demand. An evaluation period of seven years also gives some indication about short-term and medium-term "performance" of the alternative approaches.

The chapter continues with a brief overview of recent demand parameters and developments for dairy consumption in the Netherlands. Some concluding remarks are added at the close of the chapter.

SPECIFICATION OF DEMAND EQUATIONS

Starting from general static demand theory (Theil 1975, 2) per capita demand for a product is determined by per capita disposable income and prices of all products (or product groups). Using the price index of total consumer expenditure as a denominator for all price variables and the income variable, and assuming the absence of "money illusion," the demand function is

$$C_i = f(P_1, \ldots, P_i, \ldots, P_n, Y) \tag{11.1}$$

where C_i is per capita demand for product i, P_i is real price of product i ($i = 1, \ldots, n$), and Y is real total consumer expenditure per capita.

It is often reasonable to assume that the consumption of one product is only slightly influenced by the real prices of many other products (Deaton and Muellbauer 1980) for approximation, and only the clearly complementary products or substitutes have to remain in the demand equation (11.1). Moreover, consumption is often influenced more by changes in consumer habits, family structure, medical opinion, and such than by changes in prices and income. Therefore, a time element is introduced in the specification to represent the change in consumer habits. Furthermore, it has been assumed that only one price in addition to the own-price influences the consumption of a particular product to yield

$$C_{i,t} = f(P_{i,t}, P_{j,t}, Y_t, T) \tag{11.2}$$

where t represents a time subscript and T a time-trend variable.

The exact functional specification of a demand equation is often a matter of custom. However, applying general rules about specification of demand functions, explained below, the following result can be obtained:

$$C_{i,t} = a_0 + a_1 P_{i,t} + a_2 P_{j,t} + a_3 \ln Y_t + a_4 T (+ a_5 DT) + u_t \tag{11.3}$$

where a_k ($k = 0,...,5$) are unknown demand parameters and u_t is a disturbance at time t. Changes in the shifts of consumer habits can be investigated by adding a dummy trend (DT) variable to equation (11.3).

The demand equation in (11.3) has the following properties:

1. Demand becomes more price elastic for an increasing own, real price. This can be easily illustrated by simplifying equation (11.3) to

$$C = b_0 + b_1 x \tag{11.4}$$

where C is quantity and x is price. From (11.4), the price elasticity of demand (e) is

$$e = \frac{dC}{dx} \cdot \frac{x}{C} = \frac{b_1 x}{b_0 + b_1 x} \tag{11.5}$$

Normally, an own-price elasticity is negative, thus $b_1 < 0$ and $b_0 > 0$; then for $x_2 > x_1$

$$\left| \frac{b_1 x_2}{b_0 + b_1 x_2} \right| > \left| \frac{b_1 x_1}{b_0 + b_1 x_1} \right| \tag{11.6}$$

Hence, the absolute value of the price elasticity of demand is higher for a higher real price.

2. For a substitute product, the cross-price elasticity of demand is positive and for $x_2 > x_1$

$$\frac{b_1 x_2}{b_0 + b_1 x_2} > \frac{b_1 x_1}{b_0 + b_1 x_1} \tag{11.7}$$

The specification implies a higher cross-price elasticity for increasing prices of the substitute product when the cross-price elasticity is smaller than 1.

3. For a complementary product (where e is negative), just as for the own-price, more effect is shown at a higher real price level [see (11.5) and

(11.6)].

4. The income elasticity of demand (e) derived from (11.3) is

$$e = \frac{a_3}{C} \tag{11.8}$$

Assuming a positive income coefficient, the income elasticity is smaller at higher real income levels.

5. If shifts in consumer habits are reflected in changes in per capita consumption, the dummy trend variable can set a new level of change. A dummy trend variable is only introduced for a clear change of trend, which is pretested. A change of trend also follows from the extinction of a previous change in habits or tastes.

These properties reflect expected consumer behavior for food products and, more particularly, for dairy products. Preliminary investigation of the implications of the algebraic specification is an important step for empirical demand analysis. Often researchers operate with a number of specifications (Boddez and Ernens 1980; Pitts 1981) and choose among them on the basis of empirical performance. This approach can be dangerous, especially in time-series analysis with a limited number of observations.

TYPES OF PRIOR INFORMATION

Although researchers may not show it clearly, they make use of several types of prior information. Often this is hidden by a very deductive way of presenting the results. If a particular approach will not give the expected results and has been replaced by another, only the successful one will be presented.

We argue for a more clear incorporation and acknowledgment of prior information. Some inclusions of prior information that are very common include (1) a general demand system as reference, (2) the algebraic specification, (3) available knowledge on the range of the parameters, and (4) adjusting predictions.

1. We started in the last section with a reference demand system (Theil 1975). Already a choice has been made about the general approach. From the literature (Deaton and Muellbauer 1980), choices of demand systems imply restrictions (see also Chapters 1 and 4).
2. The second type of prior information is imposed by the algebraic specification of the demand equation. Here common knowledge about the behavior of consumers was incorporated, but not every researcher

will use the same information and come up with an identical specification. However, a specific approach for individual demand equations is also advised by Deaton and Muellbauer (1980, 79).

3. Most important, we likely already know a lot about the particular demand function. Researchers quite often find results that are not in accordance with their expectations. This could imply that they stop research at this point because no plausible results are available. However, it is more likely that tenacious researchers will change their specifications to develop results that look reasonable. Exactly this type of prior information (e.g., information about the plausible range of parameter values) could be incorporated in the estimation procedure.
4. Before an estimated demand equation (or system of demand equations) will be used for prediction purposes, it is often adjusted for structural change. Moreover, predictions are often adapted to more plausible values by market specialists.

Each type of prior information has its merit. But we will give most attention to the third type.

Incorporation of Prior Information in the Estimation Procedure

The use of the Bayesian method in academic circles has been stimulated by Raiffa and Schlaifer (1961) and Zellner (1971) among others. However, the method is often presented so that only those with a sound statistical background understand it. Moreover, the most important part for the empirical researcher is how to actually specify his/her prior information and systematically incorporate it. This part usually receives very limited discussion.

We explain the Bayesian method in a practical way. A full description of the method has been given in Raiffa and Schlaifer (1961, Chap. 13.5), Zellner (1971, Chap. 3.2), and Judge et al. (1985, Chap. 4).

Consider the general regression model

$$y = X\beta + u \qquad (11.9)$$

where y is a vector of n observations on the dependent variable, X is a ($n \times k$) matrix of observations on the k independent variables, β is a vector of k parameters, u is a vector of n disturbances (for the elements of the vector u, we assume a normal distribution with mean 0 and variance σ^2 and no serial correlation).

In the joint probability density function for the elements of the vector of observations, y is (for given X, β, and σ)

$$\ell (y \mid X,\beta,\sigma) = \frac{1}{(2\pi\sigma^2)^{n/2}} \exp \left\{ -\frac{1}{2\sigma^2} (y - X\beta)' (y - X\beta) \right\} \tag{11.10}$$

Equation (11.10) is called the likelihood function of the observations on the independent variable. In a classical regression procedure, we derive the estimators b of β and s^2 of σ^2 using a maximum likelihood procedure equivalent in this case to ordinary least squares.

In a Bayesian procedure we incorporate our prior information on the parameters. Suppose prior ideas exist about the elements of β. These prior ideas are not exact; if they were, we could give at least one of the parameters a particular value and redefine the regression equation (11.9). Instead, we combine the likelihood function (11.10) with the a priori information using a Bayesian procedure. With respect to this procedure Zellner (1971, 13–14) states:

> Let $p(y,\theta)$ denote the joint probability density function (*pdf*) for a random observation vector y and a parameter vector θ, also considered random. The parameter vector θ may have its elements, coefficients of a model, variances and covariances of disturbance terms, and so on. Then, according to usual operations with *pdf*s, we have
>
> $$\begin{aligned} p(y,\theta) &= p(y \mid \theta)p(\theta) \\ &= p(\theta \mid y)p(y) \end{aligned} \tag{11.11}$$
>
> and thus
>
> $$p(\theta \mid y) = \frac{p(\theta)p(y \mid \theta)}{p(y)} \tag{11.12}$$
>
> with $p(y) \neq 0$. We can write this last expression as
>
> $$p(\theta \mid y) \propto p(\theta)p(y \mid \theta) \tag{11.13}$$
>
> prior *pdf* × likelihood function
>
> where $\propto$ denotes proportionality, $p(\theta \mid y)$ is the posterior *pdf* for the parameter vector θ, given the sample information y, $p(\theta)$ is the prior *pdf* for the parameter vector θ, and $p(y \mid \theta)$, viewed as a function of θ, is the well-known likelihood function.
>
> Equation (11.13) is a statement of Bayes's theorem. Note that the joint posterior *pdf*, $p(\theta \mid y)$ has all the prior and sample information incorporated in it. The prior information enters via the prior *pdf*, whereas all the sample information enters via the likelihood function.

The type of probability density function (*pdf*) assumed for specifying prior information on the parameters determines how easily the estimation problem can be handled. A natural conjugate approach is commonly used (see Judge et al. 1982, Chap. 8.3). For these distributional forms, the prior *pdf* is of the same type as the posterior *pdf* and combines easily with the likelihood function. Of course, this limits flexibility in specifying prior information.

For the function (11.10) (e.g., the product of n univariate normally distributed variables), the vector y and the matrix X form a sufficient statistic (y,X) completely determining the *pdf*.

Now write the likelihood function (11.10) in a slightly different way, dropping the constants and giving all likelihood parameters the subscript 1.

$$\ell(y|X_1,\beta,\sigma) \propto \frac{1}{\sigma^{n_1}} \cdot \exp\left\{-\frac{1}{2\sigma^2}\left(\nu_1\sigma_1^2 + (\beta - b_1)'X_1'X_1(\beta - b_1)\right)\right\} \tag{11.14}$$

where ν_1 is the number of degrees of freedom ($n_1 - k_1$, where k_1 is the rank of $X_1'X_1$), $b_1 = (X_1'X_1)^{-1} X_1'y$ is the well-known least squares estimator of β, $s_1^2 = 1/\nu_1\ (y - X_1b_1)'(y - X_1b_1)$ is the estimator of the variance of the disturbances. Then, observe that the sufficient statistic (b_1, N_1, ν_1, s_1^2), with $N_1 = X_1'X_1$, also completely determines the likelihood function.

A statistician could observe that the combination of a prior normal *pdf* of β for given σ and a marginal *pdf* for σ of an inverted gamma form could be combined with this likelihood function to yield the same form of the posterior function. This prior *pdf* can be completely described with the sufficient statistic (b_0, N_0, ν_0, s_0^2), that is

$$p(\beta,\sigma|b_0,N_0,\nu_0,s_0) \propto \frac{1}{\sigma^{k_0}} \cdot \exp\left\{-\frac{1}{2\sigma^2}(\beta - b_0)'N_0(\beta - b_0)\right\} \frac{1}{\sigma^{\nu_0+1}} \exp\left(-\frac{\nu_0 s_0^2}{2\sigma^2}\right) \tag{11.15}$$

where b_0 is the vector with prior means for the parameters β, N_0 is a positive semidefinite symmetrical matrix with rank k_0, s_0^2 is the prior variance of the disturbances, ν_0 is the number of degrees of freedom ($n_0 = \nu_0 + k_0$). For convenience, all prior parameters have been noted with subscript 0.

The prior *pdf* differs only slightly from the likelihood function (11.14);

there the multiplying σ had the power n_1, while in (11.15), the multiplying σ has the power $\nu_0 + 1 + k_0 = n_0 + 1$.

Now it can be verified that by integration of (11.15) with respect to σ, the marginal prior *pdf* of β has a Student's t distribution with (Raiffa and Schlaifer 1961, 320, 345)

$$E(\beta) = b_0 \tag{11.16}$$

$$V(\beta) = N_0^{-1} \frac{\nu_0}{\nu_0 - 2} s_0^2 \tag{11.17}$$

where $E(\beta)$ and $V(\beta)$ represent the expectation and the variance-covariance matrix of β, respectively. Information on b_0 completely specifies (11.16). According to (11.17), with prior information about $V(\beta)$, s_0^2, and ν_0, we can derive the elements of N_0. For reasons that will be made clear later, we will not try to find N_0, but will start with $V(\beta)$ and derive N_0 by means of equation (11.17).

Combining the likelihood function (11.14) with the prior *pdf* (11.15), as in equation (11.13), leads to the posterior *pdf* in β and σ

$$p(\beta, \sigma | b_2, N_2, \nu_2, s_2) \propto \frac{1}{\sigma^{k_2}} \tag{11.18}$$

$$\exp\left\{ -\frac{1}{2\sigma^2} (\beta - b_2)' N_2 (\beta - b_2) \right\} \cdot \frac{1}{\sigma^{\nu_2 + 1}} \exp\left(\frac{\nu_2 s_2^2}{2\sigma^2} \right)$$

This posterior *pdf* (11.18) is of the same type as the prior *pdf*. The parameters of this posterior distribution are

$$N_2 = N_0 + X'X; \text{ where } k_2 \text{ is the rank of } N_2 \tag{11.19a}$$

$$b_2 = N_2^{-1} (N_0 b_0 + X'y) \tag{11.19b}$$

$$\nu_2 = (\nu_0 + k_0) + n_1 - k_2 \tag{11.19c}$$

$$s_2^2 = \frac{1}{\nu_2} \left((\nu_0 s_0^2 + b_0' N_0 b_0) + y'y - b_2' N_2 b_2 \right) \tag{11.19d}$$

We reached interesting implications of the natural conjugate approach. The parameters (n_2, b_2, ν_2, s_2^2) give a complete description of the posterior parameters (β, σ), just as (n_1, b_1, ν_1, s_1^2) give a complete description of the likelihood function. Thus, the only thing we have to do is combine the sufficient statistics of the likelihood function with the sufficient statistics of the prior *pdf*. The results for the posterior *pdf* for (β, σ) follow logically.

We are mainly interested in β and in the variance of the disturbance. Just like (11.16) and (11.17), we can now specify

$$E(\beta) = b_2 \tag{11.20}$$

$$V(\beta) = N_2^{-1} \frac{\nu_2}{\nu_2 - 2} s_2^2 \tag{11.21}$$

where $E(\beta)$ and $V(\beta)$ are the expectation and the variance-covariance matrix of a Student's t distribution. Moreover, s_2^2 and ν_2 characterize the random variable σ. With these convenient distributional assumptions, the introduction of prior information involves only a small expansion of the regression program.

Another observation concerns the ranks of the matrices N_0 and N_1. It is not necessary to supply prior information for all elements of β. For example, it is difficult to specify prior information on the constant of the demand equation (11.3). And information on changes in consumer habits also may be limited. Prior information for only a subset of β can be specified, filling in b_0 and N_0 at the other places with zeros. Even more important is the fact that $N_1 = X_1'X_1$ does not have to be of full rank. In this circumstance one cannot determine b_1, but in equations (11.19a–d), nowhere has b_1 been used. The only necessary condition is that N_2 is of full rank.

These observations imply that (1) one can even add a "new" variable to the demand equation when prior information is available; and (2) the total number of observations can be smaller than the number of variables in the demand equation. Of course, the latter (2) makes it impossible to compare some of the likelihood parameters with the posterior parameters.

Both observations are useful in situations where new products have been developed or when market structure has changed drastically.

Formulating the A Priori Information

Prior information about a demand parameter can come from different sources (see also Chapters 8 and 9). Examples are listed below.

1. Consumer panel data can be or have been used for (a) time-series analysis of price elasticities and also for determining income elasticities and shifts in consumer behavior; (b) cross-sectional analysis for income elasticities; and (c) specific analysis for different social groups.
2. Results derived from general demand systems, which deal mostly with consumer expenditure and product prices. These demand systems

usually have very broad product definitions.
3. Results from empirical analysis in other countries or areas with a similar behavior of consumers.
4. Information from established consumer theory.
5. Other information relevant for the determination of the final parameter estimate.

These different types of information should be introduced preserving the sufficiency of the statistic, as discussed below.

1. The expectation of elements of the parameter-vector β [see equation (11.16)] should be taken into account. Researchers and market analysts may have reasonable ideas about the mean of the elements of β.
2. If estimation results from other sources are available, the elements of the variance-covariance matrix can be stated fairly easily. However, many researchers present only the standard deviations (or t-values) of their estimated parameters. In that case, no information about the covariance elements is available. Moreover, it is our experience that dealing with covariances is often more difficult than using intercorrelations. However, this does not solve the information problem. Therefore, our common practice is (a) increasing the standard deviations derived from prior information; the multiplying and inflation factor depending on the reliability and relevance of the prior information. The multiplication factor is smaller if covariances or intercorrelations of the β's are available; (b) only specifying intercorrelations between elements of the coefficient vector β, when information is available. Otherwise, the intercorrelations are kept at zero. From the standard deviations and intercorrelations, the matrix $V(\beta)$ in equation (11.17) is constructed.
3. Prior variance of the disturbances is a more difficult point. Raiffa and Schlaifer (1961, 344) assume that the parameters s_0^2 and v_0 of an inverted gamma-2 function, which is the right-hand part of the *pdf* in (11.15), are known. Harkema and Kloek (1969, 10) say that "prior knowledge bears more directly on the size of the disturbances themselves than on the size of their standard deviation s_0." It is our experience that neither method works. The same can be said about the specification of prior information on s_0^2 and v_0 used by Judge et al. (1982, 223–26). This is a very special case that depends on an assumption of constant returns to scale for a production function. Researchers usually have no intuitive ideas about the variance or value of disturbances. Moreover, the specification of an equation (linear, log-linear, changes, relative first differences, etc.), the time period of observation, and the covering of the market are all factors that influence the available information. According to our experience, most researchers

have an idea about the percentage of variance of the dependent variable or the coefficient of determination R^2, explained by their prior information. By definition

$$R_0^2 = 1 - s_0^2 / s_y^2 \tag{11.22}$$

$$s_0^2 = (1 - R_0^2)\, s_y^2 \tag{11.23}$$

where s_y^2 is the variance of the dependent variable y. The prior coefficient of determination (R_0^2) depends on the specification. A model for a time series of quantities of demand [like (11.3)] leads, for example, to a much higher percentage of explained variance than a model for the same variables in relative first differences. Moreover,(R_0^2) may depend on the number and type of variables for which b_0 has been specified. Dropping trend variables in the prior specification usually gives a considerable decrease of R_0^2.

4. Since v_0, the prior degrees of freedom, is the most difficult parameter in this whole procedure, we use the value 10 unless other information is available (e.g., the number of degrees of freedom in a previous estimation procedure). Of course, one can always vary this parameter to do sensitivity analysis. The influence of v_0 can be traced as follows: conditional on the specification of standard deviations for the β's and s_0^2, a smaller v_0 implies smaller standard deviations of the posterior parameters; posterior parameters become more "certain." This follows directly from equation (11.17). However, there is also another influence: a smaller v_0 influences both (11.19c) and (11.19d). The effect on v_2 is quite clear: a smaller v_0 implies a lower value for the posterior degrees of freedom (v_2). However, the influence on the posterior parameter s_2^2 heavily depends on the specification of s_0^2 and, therefore, R_0^2, in relation to the other elements of equation (11.19d). From equation (11.19d), s_2^2 decreases for a smaller v_0 when

$$s_0^2 > (b_0' N_0 b_0 + y'y - b_2' N_2 b_2)/(k_0 + n_1 - k_2) \tag{11.24}$$

which is generally right, especially for a specification where selected elements of β are considered.

Summing up these effects of v_0, it can be said that most often a smaller v_0 gives a higher certainty of the posterior parameters. One often observes that researchers use very small values of v_0, like the minimum value of 3. Unless we have other indications, we prefer to work with a higher number.

The specification of both s_0^2 and v_0 also influences the uncertainty of the prediction.

With this elaborated discussion of the four elements of the sufficient statistic, we have defined a workable method for the applied researcher. A complete numerical example using this specification of prior parameters have been given in Oskam and Osinga (1982, 401–2). However, different approaches for the specification of prior information exist.

Other Informative Priors

The informative natural conjugate prior dealt with above can be considered as the result of a first-round estimation. Say there are time-series data available for a period of 20 years. All observations can be used for a regression analysis. However, it is also possible to divide the observations into two groups and to estimate regression equations for both groups. If the results of the first regression are used as prior information for the second regression, the result is identical to the regression for the total period (Zellner 1971, 70–72). This procedure indicates another way of generating prior information: let market specialists make conditional predictions of demand, for example, vary independent variables such as prices and income, and let them predict market demand.

From a regression on these predictions the sufficient statistics of a prior probability density function can be derived: the predicted prior. This method has been applied and advocated by Winkler (1967, 1980) and Kadane et al. (1980). Our experience with this method was not positive (Oskam 1974). For a market where a good empirically tested model was available, specialists predicted quite far from the particular model. However, we have not used the interactive method of prediction, where market specialists are confronted with the implications of their first-round predictions.

A special case of an informative prior that is easy to specify has been suggested by Zellner (1982): the *g*-prior. In this case the prior information only consists of expectations for the parameters in equation (11.3) [see also equation (11.16)]. The matrix N_0 in equation (11.17), however, is directly related to the matrix N_1 by means of $N_0 = g_0 N_1$, where g_0 has a value greater than zero. This method can also incorporate diffuse prior information on the variance of the disturbances. This means roughly that the researcher has no idea about the size of these disturbances. In this case the posterior parameter vector β is a simple weighted average of the prior parameter-vector and the likelihood parameters (see Judge et al. 1985, 110–11)

$$b_2 = (g_0 b_0 + b_1)/(1 + g_0). \tag{11.25}$$

Although this prior is attractive because of its simplicity, it has a serious drawback. If multicollinearity is a problem, then this approach will not cure the problem. Because multicollinearity is a major problem in time-series analysis with a limited number of observations, we are always looking for independent information about price elasticities, income elasticities, and trends. Beyond that, the method cannot be applied when the number of observations is lower than the number of explanatory variables.

Another type of prior information, resulting in so-called poly-t-densities (see Judge et al. 1985, 111–12), seems to have good properties. We have no experience with these types of specifications for prior information. Moreover, the resulting posterior probability density function is less easy to interpret for an inexperienced person than the results from the natural conjugate method.

TWO TEST STATISTICS

Researchers and market specialists are often interested in the question: Is there a serious difference between the prior information and the likelihood function? This can be tested by using the following statistic (see also Judge et al. 1982, 189–99)

$$F\text{-prior} = \frac{\text{Restricted sum of squared errors} - k_1 \cdot s_1^2}{k_0 s_1^2} \sim F_{v_1}^{k_0} \qquad (11.26)$$

where "restricted sum of squared errors" is the sum of squared disturbances, which follow from an estimation procedure where the specified elements of the prior parameters (b_0) are used as values in the maximum likelihood estimation. Equation (11.14) gives k_1,v_1, and s_1^2; k_0 is in equation (11.15). Because this is an F-statistic, one can easily evaluate whether the prior parameters are significantly different from those implied by the data. When there is a significant difference, it is difficult to decide what to do.

A second statistic we often use is the posterior determination coefficient

$$R_{pos}^2 = 1 - (y - Xb_2)'(y - Xb_2) / \sum_{i=1}^{n_1} (y_i - \bar{y})^2 \qquad (11.27)$$

where $\bar{y} = \frac{1}{n_1} \sum_{i=1}^{n_1} y_i$, with n_1 = the numbers of observations (see equation 11.14), y and X are explained under equation (11.9), and b_2 has been defined in equation (11.19b). The share of explained variance when the posterior mean values b_2 are used is given by R^2_{pos}. Of course, R^2_{pos} is always lower than R^2, the determination coefficient of the likelihood parameters b_1. The difference between the two measures indicates the loss in fit when the prior information is used.

PREDICTION IN A BAYESIAN CONTEXT

With the posterior parameters, prediction is straightforward (Zellner 1971, 72–75). A number of q predictions generates the following probability density function (*pdf*) for the predictions

$$\tilde{y} = \tilde{X} b_2 + \tilde{u} \tag{11.28}$$

where $\tilde{X}$ is the $q \times k$ matrix of given values of the independent variables in the q future periods, $\tilde{y}$ are the predicted values, $\tilde{u}$ are the q future disturbances, mostly assumed to have a normal distribution with mean zero, common variance and no intercorrelation, and b_2 has been defined in equation (11.19b).

The expected value of the variable to be predicted is

$$E(\tilde{y}) = \tilde{X} b_2 \tag{11.29}$$

while the variance-covariance matrix $V(\tilde{y})$ is

$$V(\tilde{y}) = \frac{v_2 s_2^2}{v_2 - 2} (I + \tilde{X} N_2^{-1} \tilde{X}'). \tag{11.30}$$

The predicted values originate from a multivariate Student's t density. Because the variance-covariance matrix is, in general, not a diagonal matrix, future predictions will be dependent.

It is also possible to work with disturbances that are interrelated and have nonzero means. In that case the matrix I needs to be modified.

Notice that prediction in a Bayesian context means the determination of a *pdf* for the variable to be predicted. This gives the possibility of making different types of risk analysis afterwards.

We can now offer insight into the effects of s_0^2 and v_0 on the uncertainty of the prediction. Higher values for s_0^2 and v_0 increase s_2^2 (see equation 11.19d) and will increase the uncertainty of the prediction. The effect of a higher v_0 on the quotient $v_2/(v_2 - 2)$ is mostly less important. However, a higher value of s_0^2 and v_0 also influences N_0 in equation (11.17) and this affects the matrix N_2^{-1} in equation (11.30) so that higher s_0^2 make predictions become more certain, while higher v_0 makes them more uncertain. In practice, a higher v_0 always increases the uncertainty, but the first effect of a higher s_0^2 (e.g., increase of uncertainty) will usually be more important.

General Scheme for Informative Prediction

Prediction of demand, based on the above mentioned method, proceeds according to Figure 11.1. This figure illustrates several types of information influencing the predictions of demand. The "additional information about future developments" refers to a possible adjustment of demand equations before deriving the predictions and also to possible adjustment of predictions afterwards.

The scheme illustrates that the general philosophy of a Bayesian approach is different from classical econometric methods. All relevant information is incorporated to reach the final result: a probability density function of the variable to be predicted. Although we prefer to incorporate more objective information, more subjective forms of information, like the ideas of the decision maker, can easily be used. Of course, for such a system to operate efficiently relevant computer programs that are easy to run after a revision of data must be used. Most computer packages such as SPSS, SAS, TSP, and so on do not contain these types of programs. An overview of available programs, mainly developed in the United States, has been given by Press (1980). We use a program that is part of an own-developed data base system: ADBS.

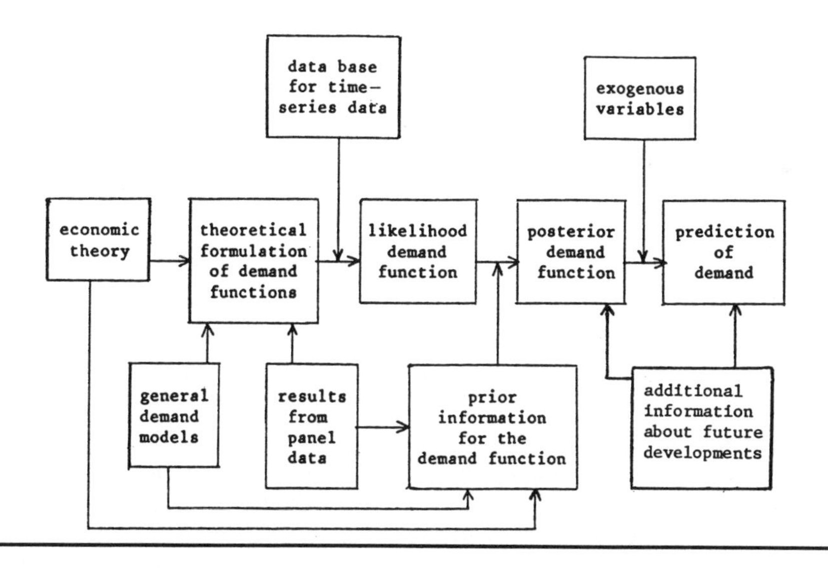

Figure 11.1. Complete system for prediction of demand

A COMPARISON OF ALTERNATIVE FORECASTING METHODS

In a previous study we derived a number of demand functions for dairy products in the Netherlands. In the study a Bayesian approach was used and the functions were estimated for the period 1965–79 (Oskam and Osinga 1982, 370, 371). We concluded that the likelihood estimates were very unreliable (because of multicollinearity), while the results of posterior analysis were very acceptable. Moreover, the relatively vague prior information was greatly improved by the addition of extra information from data (p. 375).

We now investigate the question of the accuracy of predictions generated by different methods: How do the results of normal econometric analysis compare with simple methods of prediction based on trend estimation? How do the results of the Bayesian method compare with classical econometric work? Ex post predictions for the period 1980–86, based on the estimated demand equations over the period 1965–79, were evaluated to answer these questions. A Theil coefficient was used to evaluate the predictions.

$$T = \left\{ \sum_{i=1}^{n} (P_i - A_i)^2 \,/\, \sum_{i=1}^{n} A_i^2 \right\}^{1/2} \qquad (11.31)$$

where P = predicted mean value, A = actual observation, and n = number of observations.

This statistic gives the per unit difference between predicted and actual values; thus, a Theil coefficient of 0.03 implies a difference of 3 percent between the values. Of course, other prediction measures such as turning point criteria can be used for evaluating performance. However, the Theil coefficient is a good measure for general information. Besides that, we will investigate the share of predictions outside the 90 percent reliability interval.

Results for a Trend Function and Two Different Estimation Methods

These functions were estimated using a linear trend function. A dummy trend variable [see also equation (11.3)] was used for products that had an important change in trend after 1973. This made the trend model comparable with the "normal" demand equation (11.3). The likelihood function was estimated using standard regression methods. There are some slight differences between the original published likelihood functions and those presented, because of a data revision. Also, two variables that had apparently no influence were dropped, the price of coffee in the demand equation for condensed milk, and the price of meat products in the demand equation for cheese. The results of the Bayesian method incorporate the prior information available at that time (see Oskam and Osinga 1982, 370, 371).

The prediction performance over the period 1980–86 and for all dairy products has been summarized in Table 11.1. If the difference between the prediction with a trend model and the actual value is fixed at 100 percent, a prediction based on the likelihood function gives (with ex post data for the variables of the model) a difference of 110 percent and a Bayesian method leads to a difference of 70 percent.

Clearly, the predictions from the likelihood function are poorer than the trend predictions. This implies that using predictions of a normal type of demand analysis has two disadvantages: it performs worse, and it is more work. Above that, the ex ante predictions can be more out of line. Only the Bayesian method performs here better than the trend model, although one has to keep in mind that it is an ex post Bayesian prediction.

However, such conclusions also depend on the terms of the predictions (see Table 11.2). For the short term, trend predictions and to a lesser extent predictions based on the likelihood function show the best results, while for a longer term, the Bayesian method is superior. Moreover, even for the Bayesian method it is questionable if ex ante predictions would give better results. All depends on a good prediction of the explanatory variables in the demand model, a critical factor in use of such models in forward-looking decision contexts.

Table 11.1. Average Theil coefficients for the prediction of per capita consumption dairy products in the Netherlands, 1980–1986

Product	Bayesian method	Likelihood function	Trend model	Weight (share)
Consumption milk and milk products	0.0444	0.0438	0.0471	0.343
Cream	0.0638	0.1306	0.2078	0.046
Condensed milk	0.1115	0.1208	0.0330	0.076
Cheese	0.0345	0.1269	0.0290	0.379
Butter	0.0848	0.0343	0.2330	0.156
Weight average Theil coefficient	0.053	0.084	0.076	
Theil coefficient in percentage of trend model	70	110	100	

Table 11.2. Annual Theil coefficients for the weighted average of dairy consumption in the Netherlands

Year	Bayesian method	Likelihood function	Trend model
1980	0.028	0.027	0.019
1981	0.039	0.040	0.038
1982	0.054	0.045	0.060
1983	0.049	0.080	0.085
1984	0.057	0.092	0.072
1985	0.051	0.114	0.102
1886	0.053	0.113	0.093

Note: For weights see Table 11.1.

These conclusions are less clear for the summed differences between actual and predicted consumption of dairy products (see Table 11.3). These summed differences are in kg of milk equivalent units. Especially for the trend method opposite differences cancel out. Such an interpretation of differences between actual level and prediction, however, is only appropriate when one is interested in aggregate demand. Questions about future demand for particular products, which are important for investment decisions for example, can only be answered adequately with demand functions for individual products. But even at the aggregate level the Bayesian method gives better ex post results at a term of seven years.

Table 11.3. Actual and predicted per capita consumption of dairy products in the Netherlands in 1986 (kg per capita)

Product	Actual	Bayesian method	Likelihood function	Trend model	Weight[a]
Consumption of milk and milk products	131.02	124.14	139.10	123.69	0.75
Cream	2.85	3.08	3.38	3.67	4.9
Condensed milk	9.81	8.13	7.99	9.27	2.2
Cheese	14.52	14.14	16.87	14.85	8.1
Butter	4.05	3.83	4.25	5.29	12.0
Total in milk equivalent units	300.0	286.6	326.1	314.9	
Difference from actual consumption (%)	—	−4.5	+8.7	+5.0	

[a] Conversion factors for computing milk equivalents.

We also used an F test to determine whether the prior information was significantly different from the data [see equation (11.26)]. Test results and prediction performance are presented in Table 11.4. It is noteworthy that for both cream and cheese there was a significant difference between the prior information and the data, while for these products the Bayesian method performed best. We always check our prior information, but if it is statistically different from the sample information there seems to be no reason to drop the prior information.

Table 11.4. Significance level for a F test on the difference between prior information and time-series data, together with prediction performance

Product	Significance level F-prior	Best prediction by Bayesian method/likelihood
Milk and milk products	0.26	nearly equal
Cream	0.016	Bayesian method
Condensed milk	0.85	nearly equal
Cheese	0.002	Bayesian method
Butter	0.91	likelihood

The last point that we investigated was the number of predictions outside the 90 percent prediction interval or reliability interval. For a total of 35 predictions, a number of 3 or 4 (e.g., 10 percent) would be normal. The Bayesian method had 6 (17 percent), the likelihood function 9 (26 percent), and the trend function 13 (37 percent). One may conclude that the trend model especially underestimated future uncertainty, because a trend was too simple a model and there were "structural changes" between the estimation period and the prediction period.

Of course, this is only one example where a Bayesian approach generates best medium-term predictions. However, in an earlier investigation, we derived the same results for the United Kingdom. The only difference was that for the United Kingdom the likelihood function performed better than the trend approach (Oskam 1986).

CHARACTERISTICS OF DAIRY CONSUMPTION IN THE NETHERLANDS

Reestimation of the demand equations over the period 1965–86 gives more up-to-date insight into the characteristics of demand for dairy products. Before we estimated these equations, the prior information was changed slightly (see Oskam and Osinga 1982, 370, 371 for details). The prior price elasticity of demand for butter was decreased from −1.5 to −1.0

because butter is becoming less of a luxury product. Also, the prior coefficients for the quantity coefficients of Christmas butter (second grade butter) and special arrangement butter were dropped. No recent information was available and we decided that the coefficients could best be derived from the data. Moreover, a dummy trend variable was added to the demand equation for condensed milk, starting in 1980. Apparently there was a shift in the trend for the period 1980–86.

The results for this estimation, again using the Bayesian method, are shown in Table 11.5. Because these five products cover about 85 percent of total human consumption of dairy products in the Netherlands, they give a nearly complete picture. Most noteworthy is the positive trend of about 0.65 percent over the period 1980–86.

Over the periods 1965–73 and 1974–79, we observed trends of about −2.5 percent and +0.03 percent (Oskam and Osinga 1982, 396). This implies that current total dairy consumption shows a clear upward trend after a very drastic downward shift in the period 1965–73. Causes are not investigated, but they could be stabilization in retailing after drastic changes in the 1970s; product development by the dairy industry; more attention by the dairy industry to advertising and promotion in the domestic market, also due to large surpluses and stronger competition between firms/cooperatives; and no additional negative effect on the consumption of dairy products for health reasons.

Assuming a constant real price for dairy products, an increase of real consumer expenditure of 1 percent per year, and an increase of population of 0.4 percent per year, the figures indicate a growth rate of total consumption of 1.3 percent per year. However, one has to keep in mind that every year the quantity of dairy products produced from a particular quantity of milk shows a slight increase.

Table 11.5. Characteristics of dairy consumption in the Netherlands

Product	Price elasticity	Income elasticity	Trend[a] (%)	Weight[b]
Milk and milk products	−0.14 (0.11)[c]	−0.14 (0.07)[c]	−0.67	0.328
Cream	−0.31 (0.22)	1.08 (0.26)	1.31	0.047
Condensed milk	−0.16 (0.09)	0.20 (0.11)	0.25	0.072
Cheese	−0.34 (0.13)	0.36 (0.14)	1.78	0.392
Butter	−1.00 (0.22)	0.58 (0.25)	0.56	0.162
Weighted average	−0.37 (0.07)[d]	0.25 (0.07)[d]	0.65	

[a] Calculated as percentages of the 1986 level; only for the period 1980–1986.
[b] Based on 1986 consumption figures.
[c] Standard deviation within parentheses.
[d] Based on the assumption of independent standard deviations for the particular products.

CONCLUSION

Based on our experience, we see that the synthesis of information from different sources improves analytical insights into the factors determining dairy consumption. Although short-term prediction based on trends or related methods may be preferred, for medium-term prediction and for conditional predictions a Bayesian method gives better results. Especially in a decision-making context, the Bayesian approach with probability distribution of the predictions is highly preferable to other methods. Although our approach is too aggregate for the particular questions of the dairy industry, the results are often used for policy analysis (Landheer 1987; Oskam 1989; Oskam et al. 1988). Reluctance to use Bayesian methods can be explained by the rather intricate method and the lack of adequate computer programs. We may expect improvements on these factors in the future.

REFERENCES

Boddez, G. R. and M. Ernens. 1980. *Agricultural Forecasts 1985 for EUR-6, EUR-9 and Members States*, vol. 1–9. Leuven: CLEO.

Deaton, A. and J. Muellbauer. 1980. *Economics and Consumer Behaviour*. Cambridge: Cambridge University Press.

Harkema, R. and T. Kloek. 1969. *An Example of an Informative Prior Distribution in a Decision-Making Context*. Report 6913 of the Econometric Institute, Rotterdam.

Judge, G. G., W. E. Griffiths, R. C. Hill, H. Lutkepohl, and T-C Lee. 1985. *The Theory and Practice of Econometrics*. 2nd ed. New York: Wiley.

Judge, G. G., R. C. Hill, W. E. Griffiths, H. Lutkepohl, and T-C Lee. 1982. *Introduction to the Theory and Practice of Econometrics*. New York: Wiley.

Kadane, J. B., J. M. Dickey, R. L. Winkler, W. S. Smith, and S. C. Peters. 1980. "Interative Elicitation of Opinion for a Normal Linear Model." *Journal of the American Statistical Association* 75:845–854.

Landheer, J. 1987. "The Ratio of Fat Value to Non-Fat Value of Milk: An Econometric Model of Effects of Intervention Price Changes in the EC." *European Review of Agricultural Economics* 14:161–178.

Makridakis, S., S. C. Wheelwright, and V. E. McGee. 1983. *Forecasting Methods and Applications*. 2nd ed. New York: Wiley.

Oskam, A. J. 1974. *Combining Intuitive and Statistical Information: An Experimental Investigation*. Department of Agricultural Economics, Wageningen Ag. University (mimeo).

_____. 1986. "Analysis of Demand for Dairy Products in the Netherlands and the United Kingdom, Using a Synthesis of Different Types of Information." In *Workshop on Demand Analysis and Policy Evaluation*, edited by D. P. Stonehouse. IDF Bulletin No. 197:171–194. Brussels: International Dairy Federation.

_____. 1989. "Principles of the EC Dairy Model." *European Review of Agricultural Economics* 16:463–497.

Oskam, A. J., D. D. van der Stelt-Scheele, J. Peerlings, and D. Strijker. 1988. *The Superlevy—Is There An Alternative*? Kiel: Wissenschaftverlag Vauk.

Oskam, A. J. and E. Osinga. 1982. "Analysis of Demand and Supply in the Dairy Sector of the

Netherlands." *European Review of Agricultural Economics* 9:365–413.

Pitts, E. 1981. *Factors Affecting Demand for Butter and Margarine in the European Community.* Dublin: An Foras Taluntais.

Press, S. J. 1980. "Bayesian Computer Programmes." In *Bayesian Analysis in Economics and Statistics,* edited by A. Zellner, 429–442. Amsterdam: North Holland Publishing Company.

Raiffa, H. and R. Schlaifer. 1961. *Applied Statistical Decision Theory.* Boston: Harvard University, Division of Research.

Theil, H. 1975. *Theory and Measurement of Consumer Demand.* Vol. 1. Amsterdam: North Holland Publishing Company.

Winkler, R. L. 1967. "The Assessment of Prior Distributions in Bayesian Analysis." *Journal of the American Statistical Association* 62:776–800.

_____. 1980. "Prior Information, Predictive Distributions, and Bayesian Model-Building." In *Bayesian Analysis in Econometrics and Statistics,* edited by A. Zellner, 95–109. Amsterdam: North Holland Publishing Company.

Zellner, A. 1971. *An Introduction to Bayesian Inference in Econometrics*, New York: Wiley.

_____. 1982. *On Assessing Prior Distributions and Bayesian Regression Analysis with g-Prior Disbributions.* Revised paper, presented to the Econometric Society Meeting, Denver.

CHAPTER 12

A Review of the Use of Demand Analysis for Milk and Dairy Products in the United Kingdom

R. E. Williams

DEMAND ANALYSIS should have a practical use in market management either by those responsible for policies regulating the dairy industry or by large commercial operators in the marketplace. This chapter will give examples of these uses of demand analysis in the United Kingdom. The United Kingdom is a member of the European Communities (EC), and its milk and dairy product markets are regulated by the EC's common organization of the milk sector. The United Kingdom is therefore part of the wider European market. EC policies are nevertheless frequently analyzed for their effects on different member states, and there are also special features of the British market resulting from history, which give it and the analysis of it an interest of its own.

What is demand analysis? For the purposes of this chapter I define "demand analysis" as a model-building or analytical approach to specific problems of demand. Demand analysis may be in the province of the economist alone and be concerned with the specification of demand schedules, which attempt to show the amounts likely to be bought at different levels of price, or with the relationship between income levels and the demand for a product (so-called Engel curve analysis). Economists also concern themselves increasingly with relationships between advertising/promotional expenditure and the demand for a product. Econometricians seek to measure in their "models" effects of changes in one economic variable, if others are held constant.

Demand analysis in the widest sense however should go beyond the econometrician to the market researcher. For the statistical approach to economic variables—prices, incomes, and quantities bought—can be applied to noneconomic variables, when data can be obtained through market research sources. Market researchers concern themselves with buying patterns, frequency of purchase, variations in quantity on purchase occasions, switching between types of product and/or brand loyalty, and dynamic movements of markets in terms of who buys and consumes a product and its close substitutes where and when. In the widest sense, therefore, demand analysis becomes the analysis of consumer behavior in the marketplace. Models of demand will seek to explain *why* consumers behave as they do and will not present merely the facts of that behavior, which is usually as far as market research itself goes.

MEASUREMENT OF MARKET DEMAND AND CONSUMER BEHAVIOR

The broadest measures of market demand in a given time period (month, quarter, year, and so forth) are usually total estimates of supplies moving into consumption. In most cases this is the amount of the product manufactured plus imports minus exports. (Quantities traded are particularly important for dairy products in the United Kingdom.) For storable products, stock changes between the beginning and end of the time period are also important. In the case of butter and skim milk powder, stock changes are of major importance, since as residual products in the milk supply line, changes in either milk availability for butter/skim milk powder or changes in market demand may show up first of all in stocks changes. However, stocks changes for other products (such as hard-pressed cheeses, whole milk powder, and condensed milks) are important for the calculation of market off-take and commercial management.

It is not part of the present purpose to go deeply into the nature of the data sources on consumption and demand statistics for the dairy sector except to note that a critical approach is necessary for the analyst and to underline the point that "number crunching" on demand for dairy products (no less than in any other major industry) is a very delicate art. Aggregate market statistics may be derived in various ways: from the summation of commercial transactions, from estimates of product output from the use of milk, or partially or wholly by sample survey. Methods will often be dependent on the ease of availability of data from the marketing system and be influenced by the costs involved. No statistical series should be used for analysis unless the analyst is aware of the limitations of the data and especially of the definitions that are often in the small print of footnotes. The United Kingdom's *UK Dairy Facts and Figures,* published each year by the Federation of United Kingdom Milk Marketing Boards, provides a good source of series of aggregate supplies moving into consumption. Annual data, of course, often mask variation within years and, where this is thought to be important, monthly or quarterly series may be needed.

The availability of retail price series is for some products dependent on continuous survey data. The British government's national food survey and the Attwood and Audits of Great Britain Ltd. (AGB) household panels are sources most commonly used in industry circles (the Attwood and AGB data have been made available to researchers by permission of the companies through subscribers). Price series are also available from the detailed collection of prices for the calculation of the index of retail prices by the Central Statistical Office. The great benefit however of panel and continuous survey data on household purchases is that the problem of matching the definition of the price and quantity series is eliminated.

There are many well-known problems in the use of panel and continuous data for the nature and accuracy of recording prices and expenditure. A literature exists on the problems in general, but dairy products pose special problems. Liquid milk, for example, is one of the very few products purchased daily (or almost daily) by the vast majority of households. The effect of holidays or movement of individuals into and out of households for short periods will therefore be detected in panel recording, but often not in a random survey operation. The random sampling technique is likely to indicate a higher level of household milk consumption than a panel. The same might be said for other dairy products, but the effect is less when frequency of purchase is less. Butter purchases are known to be sensitive to the act of recording, because consumers may often prefer not to admit to purchasing substitutes. Cheese purchases are sometimes difficult to record accurately simply because of the way some cheese is sold (i.e., cut from the block).

Individual market demand is affected by age of consumer and family structure, and both household panel and random survey data are usually able to provide analyses of these features. This may provide a bridge between the application of an economic theory, which is intended to apply to the individual, and the analysis of market demand. The "buying unit" for most dairy products is the household. Panel data therefore, while not without limitations, can be an exceedingly rich source of material for the analysis of market demand. Buying patterns, frequency of purchase, and brand loyalty can also be analyzed from these data to give material for planning marketing strategy. Nevertheless, household panel and survey data do not cover the out-of-home or nonhousehold market, which for all dairy products is an important sector of the total market. Catering is by far the largest part of the out-of-home market, but it presents an extremely wide spectrum—from eating and drinking at work, to eating and drinking during travel, to eating out on holidays. All of these market sectors are important, but it has to be admitted that they are underanalyzed and data are difficult to obtain. As outlets, of course, they are covered by aggregate market off-take statistics from which their size can be deduced when there is sufficient confidence in the projected level of household sales using panel/survey data.

UNDERSTANDING MARKETS FOR MILK AND MILK PRODUCTS

In this section examples of demand analysis will be given for each of the main markets and, to put these examples in perspective, a description of each market and recent market developments will be given. Later sections will tackle specific uses of analysis for policy and marketing problems. A prime use of demand analysis is to give a background understanding of movements in the marketplace.

FLUID MILK MARKET

Fluid or liquid milk consumption is relatively high in the United Kingdom. Annual consumption of liquid milk in all forms (whole milk, semiskimmed, and skimmed milk) was 126.7 kg per capita compared with an EC average of 91.2 kg per capita in 1985 (Milk Marketing Board 1987). Liquid milk therefore is a significant part of consumer food budget in the United Kingdom.

There has been a gentle downward trend in consumption since the high point in the mid-1970s, when the Labor government (1974–79) placed a very heavy subsidy on milk (and smaller subsidies on certain other foodstuffs) as one plank in a policy to check the momentum of the inflationary spiral following the first oil price hike of 1973. In recent years there have been a number of major changes affecting the market, which are worth listing. Modeling cause and effect is difficult in the circumstances of very recent changes, in which there can be little association with past experience. Key features and influences have been as follows.

1. The maximum retail price of milk in the United Kingdom was controlled by the government for 45 years from 1939 to 1984. The government lifted control on January 1, 1985.
2. In the last year or so shop sales in England and Wales have begun to increase to the point where at the end of 1986 they constituted 20 percent of household sales. This has meant a contraction in household delivery, which has taken the weight of the downward trend as well as the switch to shops.
3. Within the shop trade, multiples have been battling for market share with the result that shop prices have become significantly lower than prices delivered to the doorstep. Actual retail prices paid by consumers

have begun to vary more than they did throughout most of the period of government control of the maximum retail prices.

4. There has been an increasing public consciousness of the connection between diet and health. This has been an important factor in the sharp growth of low-fat milk sales. EC legislation has enabled semiskimmed milk to be sold, whereas previous British legislation did not.
5. In the effort to maintain the competitive position of delivery rounds, low-fat milks have been offered increasingly through the doorstep service sometimes at lower prices than whole milk. Competition therefore has also been an important factor in influencing the market trend.

These factors have undoubtedly affected the response of consumers to price changes in the market. An analysis of own-price elasticities has been published by the National Food Survey each year and uses a covariance analysis of monthly data for the previous five years. Results (with standard errors in parentheses) are shown in Table 12.1. Clearly the own-price elasticity of demand for whole milk has become significantly larger in the last few years, and it is reasonable to assume that this is due to increased choice through the possibility of low-fat milk as an alternative to whole milk.

Milk consumption per capita has moved downward in the last decade by around 12 percent. Most of this fall occurred between 1975 and the early 1980s when the *real* price was increased by around 35 percent because of the move from a position of very heavy subsidy to no subsidy. With a price elasticity in most of these years of around 0.25 the price rise would have accounted for a fall in consumption of around 9 percent. Between 1981 and 1985 the *real* price of milk was stable, but consumption per capita continued

Table 12.1. Own-price elasticities of demand for liquid whole milk

Years (inclusive)	Elasticity	Standard error
1971–1975	−0.15	(0.04)
1975–1979	−0.16	(0.04)
1977–1981	−0.24	(0.11)
1978–1982	−0.24	(0.13)
1979–1983	−0.30	(0.17)
1980–1984	−0.50	(0.17)
1981–1985	−0.45	(0.20)

Source: National Food Survey Annual Reports (various issues).

Note: The results quoted can be found in the NFS annual report relating to the last year of the period of analysis in this table and Tables 12.2, 12.3, and 12.4.

to fall at an average of just over 0.5 percent per year. The analysis therefore suggests that the demand schedule for milk has been fairly stable across the last decade but may have moved to the left a little in the last few years. The shape of the demand schedule has also changed slightly and become a little more concave to the origin with the advent of low-fat milks.

Another interesting feature of the analysis of demand is that the income elasticities of expenditure and quantity purchased for liquid milk in the United Kingdom are not significantly different from zero. Analysis in some years has shown them to be just positive and in other years negative. Thus expenditure on milk and the quantity bought are not influenced by changes in the level of real incomes (National Food Survey Committee 1974–1987).

There is a contrast between the income elasticity of expenditure on milk and that on food as a whole. The income elasticity of expenditure on food as a whole has been around 0.20 during the period 1976–1985. This may be interpreted to mean that a 10 percent increase in real income will raise food expenditure by around 2 percent. As incomes rise therefore, food expenditure (other things being equal) will tend to rise. The results of the analysis appear in Table 12.2.

Food expenditure rises slightly with rising incomes, mainly because expenditure on "convenience"-type or prepared foods rises but also because consumers may switch demand to higher quality and more expensive types of food. Milk demand does not benefit from these changes. With the milk expenditure elasticity below that of all food, it might be expected (again, other things being equal) that milk expenditure would take a declining share of the consumer's food budget over time. There may be some tendency for this to happen, but it is certainly not clear-cut in the National Food Survey data because of irregular movements. These irregular movements may occur

Table 12.2. Income elasticities of expenditure on whole milk and all food

Year	Whole milk	Food	Milk expenditure as percentage of all food
1976	0.00	0.15	9.25
1977	0.06	0.10	9.50
1978	0.02	0.14	9.71
1979	0.03	0.21	9.42
1980	0.05	0.24	8.10
1981	0.00	0.25	7.85
1982	0.03	0.21	9.56
1983	0.02	0.22	9.01
1984	0.03	0.21	8.42
1985	−0.06	0.16	7.88

Source: National Food Survey Committee Annual Reports.

when other things are not equal, for example, when there is a change in the relative price of milk between years through a change in subsidy policies or market forces or, as in recent years, a switch to low-fat milks as an alternative to whole milk. The income elasticities, moreover, are calculated by cross-sectional analysis, that is, comparing households with lower and higher incomes and assuming that such elasticities applied to real income changes over time can only be approximated.

Cream Market

The cream market in the United Kingdom is interesting from an analytical viewpoint, partly because some of its features are very different from other main consumer dairy product markets. The fresh cream market is a narrow one in terms of the proportion of households that buy cream and especially the proportion that buy regularly in any week or four-week period. It is a relatively small market as an outlet for milk (compared with liquid milk, butter, and cheese) and at the same time highly seasonal. Demand for cream reaches a peak in the summer during the soft fruit season (early June to the end of July) and reaches another very high, short, sharp peak in the week before Christmas.

The fresh cream market grew rapidly during the 1960s and early 1970s, but growth has become slower in more recent years. A substantial part of the cream market is not in the form of direct sales to households, but rather for cream cakes and other prepared foods such as frozen cakes, desserts, and ice cream.

Own-price elasticity estimates for cream, published by the National Food Survey, have been more intermittent than for liquid milk and butter, and have shown a great deal of variation ranging between elasticities of −0.23 (0.37) in 1977 and 1979, to 1.06 (0.44) in 1976, and −0.95 (0.40) in 1984. The errors of estimate for cream (shown in parentheses) are high, partly because of the very narrow base of the household market. Also, there is an enormous range of types of product sold as cream with many product variations by level of butterfat, heat treatment, product pack, and size of container, none of which is standardized. These features make the market a difficult one for econometric analysis and also explain why cream is not a mass market. It is a complementary market, dependent on fresh fruit [which also has moderately high own-price elasticity of −0.53 for the years 1982 to 1987 (National Food Survey 1987)] and types of sweets eaten.

Income elasticities of demand for fresh cream are also high compared with most foods with values over 1 in most years that the calculation has been made. This is another indicator of a narrow-based market likely to have a luxury image.

Cheese Market

The cheese market in the United Kingdom is of modest size in European terms. Consumption per capita at 7.2 kg is just less than half the average for the EC as a whole. Apart from the Irish Republic, the United Kingdom has the lowest level of cheese consumption in the community (Milk Marketing Board 1987). The market has grown very slowly in the last decade, and much more slowly than the market for cheese in other EC countries.

Cheese is nevertheless a mass market in the sense that 68 percent of households buy cheese in one week and over 85 percent buy in a four-week period. Expenditure, however, constitutes only about 3 percent of the household food budget. The market is dominated by one variety, cheddar, which accounts for more than 60 percent of the market. The cheddar market of the United Kingdom is a highly competitive one with imports from a number of EC countries but primarily from Ireland and West Germany. Limited quantities are permitted under a special import levy, negotiated with the EC, from New Zealand, Australia, and Canada. There has been very low growth in the market off-take for cheddar cheese in recent years, but a struggle for share by those trading in the market.

Most of the small market growth in the cheese market as a whole has been by territorial varieties of cheese, one of the features helping to diversify the market along with vacuum packaging of factory-cut pieces of cheese sold through supermarkets. Some new cheeses have also been introduced in the medium and soft range; imports of Continental-type cheeses in this range have also tended to increase in the last decade, and consumption has tended to expand.

The evidence of the demand analysis from the National Food Survey is that the demand for cheese is fairly price-elastic compared with other foods and has tended to become more so, with values rising from around −0.20 in the first half of the 1970s to values exceeding −0.50 in the 1980s.

The difficulty however is that the analysis of the National Food Survey is for "all cheese," and the standard errors of the calculations exceed the elasticity values (Table 12.3). The significance of price as a determinant of the market for all cheese is therefore not high despite the relatively high elasticity value.

Interpretation of movements in the cheese market in price terms is less straightforward than for liquid milk. The demand for cheese has grown over the last decade by around 14 percent, measured in consumption per capita. Real prices moved up in the latter half of the 1970s, but have not done so in the 1980s.

Income elasticities for expenditure on cheese, on the other hand, have

Table 12.3. Own-price elasticities of demand for cheese

Years (inclusive)	Elasticity	Standard error
1970–1974	-0.14	(0.16)
1973–1977	-0.23	(0.28)
1976–1980	-0.37	(0.51)
1977–1981	-0.51	(0.60)
1978–1982	-0.51	(0.59)
1979–1983	-0.54	(0.74)
1980–1984	-0.54	(0.62)
1981–1985	-0.35	(0.21)

Source: National Food Survey Committee Annual Reports.

been around 0.30 in each year calculated, and the values have had very low standard errors. The demand for cheese appears likely to be significantly related to income changes and moves up with rising living standards. Cheese is a food that requires comparatively little preparation and fits in well with convenience food patterns of living and the increasing tendency for less formal meals, as life-styles change and less time is devoted to meal preparation.

Butter Market

The butter market in the United Kingdom has contracted sharply since the middle of the 1970s, and that contraction in demand has been one of the major elements in the European surplus milk problem (Milk Marketing Board 1987). Between the mid-1970s and the late 1980s the market declined by one-half, falling from 500,000 tonnes a year to 250,000 tonnes—a massive change, in food consumption terms. The question is, what light has demand analysis thrown on the reasons for this change?

Consumption of butter and margarine, taken together, has decreased in the last decade, largely because of the decline in the consumption of bread and the trend away from the consumption of farinaceous foods. Nevertheless, the decline in the consumption of "yellow fats" as a whole would account for only a small part of the decrease in butter consumption had butter not lost market share. If butter had the same share as in the mid-1970s, its market decline would have been less than 40,000 tonnes out of the total 250,000 tonnes decline.

The income elasticity of demand for butter in 1984 was 0.29. Income changes therefore have been a modest positive factor and, other things being equal, might have enhanced the market by around 15,000 tonnes. Given that margarine has a negative income elasticity, ceteris paribus, rising incomes should have enhanced butter's market share slightly in the last decade. Butter's share, however, dropped from 68 percent in 1975 to 34 percent in 1986.

There has been a great deal of analysis of the response to price changes of both butter and margarine consumption by the National Food Survey, with both own-price and cross-price elasticities calculated. Using a covariance model and removing the effects of variance between years (trend) and between months (seasonal) and calculating the coefficients from the residual variance, the committee's analysis suggests that the own-price elasticity of demand for butter has fallen quite markedly over the years (the covariance model is discussed in Brown, J.A.C 1959).

The cross-price elasticity of margarine with respect to the butter price has also declined but remains above the own-price elasticity of demand for butter (Table 12.4). The conclusion must be that the influence of the butter price itself as the main determinant of market share has declined. If an average value of elasticity were used and applied to the price movement over the decade, it would account for less than 50,000 tonnes out of butter's total decline of 250,000 tonnes. While such a simple calculation is debatable, it cannot seriously be doubted that factors other than prices have played a large part in the decline of the butter market.

The factors other than relative commodity prices that have played a substantial part in the decline in butter are the technical developments in the substitute products, the advertising and promotional weight in their marketing and the health debate. These factors have been interlinked. The appearance of "low-fat" spreads, of Krona margarine (a brand that appeared to grow in consumption particularly at the expense of butter), and more recently of "full-fat" spreads with the advantage of spreadability have undoubtedly been an important influence.

During the last decade there has been much public debate and propaganda on the influence of the consumption of animal fats on coronary heart disease. This debate has turned from time to time around the appearance in the United Kingdom of reports from government-appointed advisory committees (e.g., Committee on Medical Aspects of Food Policy

Table 12.4. Price elasticities of demand for butter and margarine

Years (inclusive)	Butter				Margarine			
	Own-price		Cross-price		Own-price		Cross-price	
1964 – 1971	−0.51	(0.08)	0.23	(0.04)	−0.29	(0.42)	0.79	(0.14)
1967 – 1974	−0.42	(0.06)	0.26	(0.04)	−0.65	(0.26)	0.77	(0.11)
1970 – 1977	−0.37	(0.06)	0.26	(0.03)	−0.52	(0.18)	0.66	(0.09)
1972 – 1979	−0.28	(0.07)	0.23	(0.04)	−0.64	(0.17)	0.55	(0.10)
1978 – 1982	−0.19	(0.10)	0.19	(0.05)	−0.32	(0.19)	0.40	(0.11)
1977 – 1984	−0.07	(0.14)	0.16	(0.07)	−0.64	(0.20)	0.31	(0.14)

Source: National Food Survey Committee.

1984). There is a widespread public belief (encouraged by the advertising campaigns for some brands) that the consumption of margarine is healthy and that butter consumption is not. The effects on butter consumption have been marked.

FORECASTING MODELS

Large organizations such as the milk marketing boards (MMB) in the United Kingdom (and no doubt other large farmer cooperatives in other countries as well) need to forecast market returns from the sale of all their members' milk in order to be able to operate their payment systems to producers. The milk marketing boards in the United Kingdom indicate to their producers provisional monthly prices for the marketing year (April through March), and these estimated prices are based on forecasts of total milk supplies and forecasts of the size of markets and returns from them. Forecasts therefore are required in this instance as part of a specific administrative task, and therefore the most important demand of the administration is to get as near to the actual outcome as possible. Explanations of market change are important only in so far as they improve forecasting performance for the near term.

Econometric models for forecasting are only useful to the extent that they can be relied on to produce accurate predictions, which implies not only that the model itself would have replicated the market behavior reasonably closely in the past, but that reasonable guesses can also be made for the independent variables for the near future. Some 60 percent or so of the income of the MMB for England and Wales comes from sales in the liquid market, and it has been found worthwhile to produce monthly estimates of the size of the market by means of an econometric model (similar problems are dealt with in Popkin 1975).

Changes have been made in models used from time to time, but a recent example of a model used to forecast the sales of full-priced milk by the MMB for England and Wales is given below. To standardize calendar month data, monthly liquid sales figures are divided by the number of days to obtain average daily sales. Because the model is short term, estimated over 36 periods, and updated twice each year, no account is taken of population or household changes because of the slow and gradual nature of shifts in these variables.

The model incorporates a time trend (which is sufficient to take up movements in taste and structural and demographic changes) and a set of dummy variables to handle the minor problems of seasonality. The model is estimated in double-logarithmic form by ordinary least squares. Attempts have been made to adjust the dependent variable with a range of income elasticities, but the value of 0.075 has been found to be optimal, although as an income elasticity it is lower than other estimates. The own-price elasticity of −0.17 is significant at the 1 percent level and agrees reasonably well with other analyses for the fluid market as a whole (whole milk and low-fat milks taken together). The model is shown in Table 12.5.

The adjusted value of $\bar{R}^2$ (for 22 degrees of freedom) indicates that 97 percent of the variation about the mean is explained. The Durbin-Watson Statistic implies that serial correlation is not a problem, and a plot of residuals confirmed this conclusion.

The level of liquid (fluid) milk sales is obviously only one (but an important) variable needed to forecast the level of market return for all milk sales for the current financial year. The above is nevertheless an example of an econometric model that can be manipulated and frequently updated on a moving basis to supply forecasts for an administrative purpose. Models have also been developed for markets other than the liquid

Table 12.5. A short-period forecasting model for liquid milk sales

Dependent variable: Average daily sales of liquid milk

Independent variables	Coefficient	t-statistic
Constant	2.00539	14.3
Trend	−0.00121	−15.8
Real price of milk	−0.17015	− 3.3
Real disposable income	0.07500	search
Seasonal dummies:		
January	−0.01360	− 3.5
February	0.01122	2.9
March	0.01476	4.3
April	0.00348	1.5
June	−0.01090	− 3.2
July	−0.01536	− 4.9
August	−0.02844	− 9.1
September	0.00348	1.5
October	0.00348	1.5
November	0.01389	4.4

Summary Statistics

$\bar{R}^2$ = 0.9662
F statistic = 64.9
Number of observations = 36
Durbin-Watson Statistic = 1.7093

Source: Milk Marketing Board.

market with varying degrees of success. The main example from the liquid market shows how demand analysis can be used successfully to provide the building blocks in an objective way for the improvement of an essential administrative task.

The improvement of forward budgeting is only one use of forecasts. Forecasts are also required for the medium and longer term for decisions on investment programs and strategic planning. Longer range forecasting invariably has to be cruder in model-building terms, usually based on structural analyses but it is nonetheless very important in the decision-making process. It may be necessary to produce such forecasts at a first stage on the assumption of no change in real prices, taking into account likely changes in real incomes, population projections, and residual trends.

At the second stage such long-term forecasts may be modified to take account of price changes that would be necessary to balance supply and demand and to check these against past trends in real prices. In the United Kingdom forecasting of this sort has also been carried out in the dairy industry and published by the National Economic Development Office (NEDO) on behalf of its subgroup, the Food and Drink Manufacturing Economic Development Committee (NEDO 1987). Forecasting of this sort might be described as descriptive analysis rather than econometric model building. This does not however reduce its importance for business planning, and it can be appropriately described as one of the uses of demand analysis.

EFFECTIVENESS OF ADVERTISING

A number of successful attempts has been made in the United Kingdom to use econometric analysis to measure the effectiveness of generic advertising on the level of liquid milk sales. At least one of these attempts (Strak and Gill 1983) sought to measure the effectiveness of generic advertising for a number of milk and dairy product markets and then to determine how a given budget might be spent across the range of milk and milk products to maximize returns to producers and processors (Strak 1983). A recent study of advertising effectiveness in the liquid market in England and Wales was commissioned by the National Dairy Council from an econometric modeling company (MMD Ltd. 1986) and some of the main results will be summarized here. A quantitative analysis of the effectiveness of advertising should not only quantify the relationship between the amount of money spent on generic advertising and liquid milk sales, but also calculate an optimum generic advertising budget for the market according to the management objective. MMD Ltd. also attempted this task as part of its work.

Whereas a model used for forecasting sales up to twelve months forward can be a short-term one that is frequently reworked with more up-to-date data, almost of necessity a model that attempts to measure the effectiveness of advertising needs to be longer term. Models that attempt to measure the effects of advertising will normally include price and income as independent variables to isolate their effect. Advertising is normally one of those factors subsumed under the trend effect in market analysis along with changes in attitudes to health and diet, product development, changes in the proportion of children and old people in the population, and changing patterns of distribution. Models incorporating a measure of generic advertising need to consider other factors influencing market trends.

There are many different ways of specifying the manner in which advertising might affect the actual levels of milk sales. Advertising might have a gradual buildup effect over a long period; it might be considered to be subject to diminishing returns; it might be most effective when above certain levels; it might be more effective in short, sharp bursts rather than holding a message steadily before the market; the effect of the message can decay as memories fade. All of these policies and ideas can be specified in model form and, indeed, were in the course of MMD's work, and many hundreds of different specifications of the model were run. Different functional forms were also tried for the model (linear, logarithmic, inverse, logarithmic-inverse, and so forth), but the semi-log form was found to be the best in the MMD Ltd. study.

There are also many ways in which generic advertising can be measured as a variable over time, and a number of these were also tried. It may be measured in terms of the *real* level of expenditure on the campaign month by month at constant prices, once a satisfactory index of the rate of inflation for the purpose is available. It may be measured in terms of the size of the television audience covered or in terms of advertising recall survey data if these are available regularly over time.

In England and Wales television ratings (TVR) were available from 1982 onwards, and recall tracking data were also available from the same date. Monthly data on the total advertising sales expenditure were available for the ten years 1975 to 1985, and for the period of overlap there was a very close association between the TVRs and the level of advertising expenditure at constant (1980) prices. Advertising expenditure in England and Wales has varied considerably in real terms over the years 1975 to 1985, and there is therefore a large variance in monthly data series, even when normal seasonal variations are removed. The substantial variations in annual data shown in Table 12.6 make this point clear.

The most successful model incorporating advertising expenditure was found, using the straight inflation-adjusted monthly expenditure, treating expenditure as a "stock" with each individual month's expenditure subject

Table 12.6. Generic advertising of milk in England and Wales at constant prices, 1975–1985 (1980 = 100)

Year	Generic advertising expenditure (1980 = 100)	Annual change
		percent
1975	55.1	—
1976	53.7	−2.6
1977	71.4	33.0
1978	77.9	9.1
1979	91.9	18.0
1980	100.0	8.8
1981	68.1	−31.9
1982	69.5	2.1
1983	62.9	−9.5
1984	52.5	−16.5
1985	63.7	21.3

Source: MMD Ltd. 1986.

to an exponential decay rate of 20 percent and added into the series for the following twelve months. In approximate terms, therefore, the effect of any particular month's expenditure is exhausted after a twelve-month period. Certain other variables were also incorporated including the child population (1–14 years), a health concern index (formulated from the answers to a number of questions in an annual survey on healthy eating), the number of press mentions of diet and coronary heart disease and dairy products, and the proportion of wholemeal bread in total bread consumption. The most successful model is given in Table 12.7.

The general conclusions drawn from the interpretation of this model were of considerable significance. First of all, the price and income elasticities indicated were close to levels that would have been expected from other models using time series. The derived elasticities from the semi-log function were −0.11 for real price and 0.14 for real earnings, and the t-statistics were well above 1.96. Second, the number of children within the population is also significant as is the health concern index. Increasing numbers of young people from a marketing viewpoint present a clear opportunity, whereas the health concern index presents a challenge to the industry (one which it is meeting by the promotion of skimmed milks in various forms). Third, generic advertising has had a small but positive effect on milk sales in England and Wales. An increase in real expenditure by 10 percent would be expected to raise milk sales by 0.1 percent. Had there been no advertising at all in 1985, the estimate is that the total market would have been 107 million liters or 1.8 percent lower than it was (National Dairy Council 1987).

Table 12.7. Analysis of the effectiveness of advertising

Dependent variable: Log of monthly average daily full-priced sales of liquid milk (1975–1985)		
Indenpendent variables	Coefficient	t-statistic
Constant	1.8958	24.8
Real earnings	0.00127	3.8
Real price of milk	−0.00116	− 5.6
Child population	0.829×10^{-4}	26.4
Health concern index	−0.0029	− 4.0
Real advertising expenditure with 20% decay rate	0.0117	3.1

Summary statistics

$\bar{R}^2 = 0.98$
Durbin-Watson Statistic = 2.15
Log-linear form

Source: MMD Ltd.

The consulting agents, MMD Ltd., made the assumption that the contribution that additional sales would make to the profitability of the industry would be 6½ pence per extra pint sold in 1985 and 1986. This might seem low. On the producer side, the additional revenue is the difference between the butter price for the milk and the liquid price, itself a difference equivalent to just over 3 pence per pint. On the distribution side it might be argued that the cost of selling the marginal pint is negligible and the whole of the distribution margin (just over half the retail price—24 pence per pint) is a contribution.

While this may exaggerate the case slightly, it makes the point that an assumed total contribution for the industry of 6½ pence per pint is low. However, based on this modest assumption, the calculations showed that the extra net profit for the industry, generated by additional sales from the advertising expenditure, was below the optimal level. The industry's total generic advertising expenditure in 1985 was £6.6 million, whereas an optimal level of expenditure would have been £7.8 million. The "extra profit" from given levels of advertising expenditure, calculated from the elasticity in the model, and the contribution assumptions are shown in Figure 12.1. The range of observations in the data used in the model is also shown, and it can be seen that the curvature of the profit line is quite small within this range. The results therefore have to be treated cautiously.

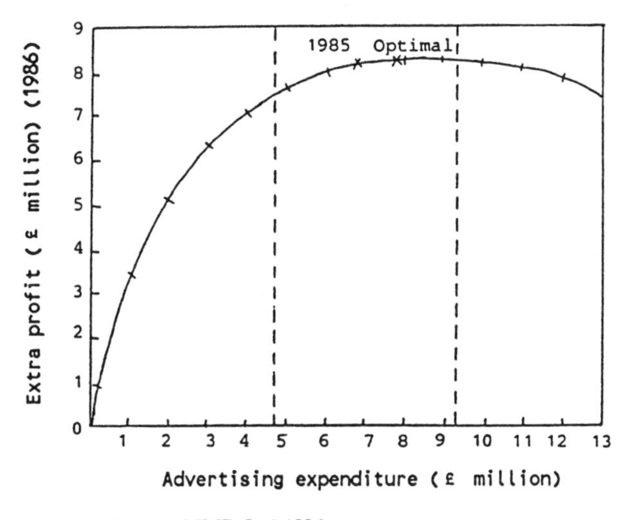

Figure 12.1. Impact of advertising on the profit of the dairy industry (Dotted lines show range that is within the experience of the model.)

OTHER COMMERCIAL PROBLEMS AND DEMAND ANALYSIS

Study of the effectiveness of generic advertising as an area of demand analysis has developed considerably in the last decade or so, and with the large sum spent this must be applauded. When the informational and analytic base on which these major decisions are taken is improved, there is at least a reasonable expectation that decisions and use of resources will improve. There are however other areas in which commercial decisions can be assisted by demand analysis. The marketing management of large organizations controls a number of policy instruments—the price of their products, promotional expenditure, quality variations—that influence the market shares for their products. In seeking to set and achieve targets for market share to maximize profits, market analysis can be a guide of considerable importance.

An example of this type of commercially oriented analysis may be taken from the British butter market. Throughout the history of policy pertaining to the British dairy industry a larger part of the butter consumed in the United Kingdom has been imported from a number of different countries.

Countries with major butter imports to the United Kingdom have established by advertising and promotional effort principal brands of butter, which might be called "flag brands." Anchor, for example, is the New Zealand brand name; Lurpak, the Danish brand name.

To compete with this situation, butter producers in the United Kingdom formed a consortium to establish a British brand, Country Life, to be nationally promoted. Not all imported or home-produced butter is sold under these principal brand labels. Some butter is sold in bulk to packers and distributors, who sell under their own labels. An important question for marketing organizations to ask, therefore, is where to set their selling prices and how much should be spent on brand advertising to increase market share consistently with maximizing profit.

Data problems for the estimation of the most completely specified models are considerable, especially for brand advertising. These limitations have to be accepted in model building. A simple model, using a double-logarithmic function relating market share of the brand to relative price of the brand compared to all other butters and a time trend, was found to give reasonably good results from a statistical point of view for a number of brands. A historic example of work carried out at the Milk Marketing Board for the three years January 1981 to the end of December 1983 (39 four-week period observations in all), using AGB household panel data, is given in Table 12.8.

Table 12.8. Market shares and relative prices for major brands in the British butter market 1981–1983

	Anchor	Country Life	Lurpak	St. Ivel
Dependent variables:				
Brand share	—	—	—	—
Independent variables:				
Constant	−1.46536	−2.12898	−1.95514	−3.36237
Relative price				
Coefficient	−8.36317	−9.67332	−3.97114	−11.5207
t-statistic	(−18.7)	(−5.8)	(−6.5)	(−6.7)
Time trend				
Coefficient	−0.00239	—	0.00987	0.01438
t-statistic	(−2.4)		(9.3)	(3.3)
$\bar{R}^2$	0.9096	0.6842	0.7489	0.6532
Durbin-Watson Statistic	1.4860	2.0589	1.0094	0.7547

Source: Milk Marketing Board, Economics Division.

In Table 12.8 the relative price coefficients can be interpreted as meaning that, for example, for Anchor butter a 1 percent increase in relative price would result in an 8.4 percent loss of market share, and for Country Life and Lurpak the loss in market share would be 9.7 and 4.0 percent respectively. This is not of course the same as a conventional own-price elasticity, but it implies conventional elasticities for brands that are well in excess of unity.

Very high elasticities for brands are an indicator that consumer loyalty to a particular brand of butter in the British market is very low. Consumers are prepared to switch from one brand to another for small price changes. The results suggest that consumer loyalty to Danish Lurpak is higher than for other butters. This may be because it is lactic butter with a distinctive taste, whereas the others are sweet-cream salted. The results also suggest that store price promotions for a brand are likely to be highly successful in the short term, but they are likely to give rise to retaliatory action, which explains why over long periods in the United Kingdom a high proportion of butter sold through the supermarket stores has been sold on promotion. Clearly, however, it is possible to use this kind of model to calculate how far price promotions are likely to pay off for one marketing organization compared with another, and at this stage the econometrician should be working very closely with the marketing manager.

This example of the use of models and their estimation for commercial marketing problems is only one among many possibilities, some of which have been probed and some have not. The British cream market, for example, is an extremely diverse one with many types of product (double, single, whipping, and half cream), different heat-treatment processes, and different sizes and types of pack. The relative pricing of these products to maximize market returns is another econometric problem. Similarly, problems can be found in the cheese market where there is also competition from imported supplies, usually clearly marked by source of origin and with a large number of types and qualities of product. The field is wide.

PUBLIC POLICY ISSUES

Governments have frequently sought to influence the level of demand for milk and milk products. The governments of the United Kingdom have for a variety of reasons. The Labor government of 1974–79, for example, placed a heavy subsidy on liquid milk as an anti-inflationary measure. All post–World War II governments in the United Kingdom had subsidies for vulnerable groups such as welfare milk for expectant mothers and children under five years (confined more recently to those on social security and the unemployed) and free or cheap milk for school children. The EC had also used subsidy policies (as has the government of the United States) to enhance demand to dispose of surplus. Governments therefore have needed to ensure that the design of programs to enhance consumption are effective in the sense that they give a good return in terms of the policy objective for the taxpayers' money. Demand analysis has a role to play, although in the nature of things it is ex post. It is surprising that the amount of published work on these aspects is not large, notwithstanding the very large sums of public money involved.

Attempts have been made to measure the effects of the general liquid milk subsidy in the United Kingdom in the 1970s (Williams 1987) and of welfare schemes in the United Kingdom (Empson 1958), and it is not proposed to summarize this material. A much more recent problem has been the European Community's attempt at internal disposal of surplus butter stocks by means of a subsidy on a limited tonnage for a limited period, the so-called Christmas butter program. How effective was the program in increasing the total level of butter consumption?

In October 1984 the EC decided to give a special subsidy on a sale of 200,000 tonnes of butter withdrawn from stocks, mostly intervention stocks. The 200,000 tonnes were allocated to member states on a key, relating approximately to national consumption, and 39,200 tonnes were allocated to the United Kingdom. The sale started at slightly different dates in member states, but in the United Kingdom is started on January 14, 1985, and was completed by the end of March 1985.

In the United Kingdom the EC subsidy was ECU 147.25 per 100 kg, which at the then "green" rate of exchange for sterling amounted to 22.8 pence per 250 g packet. The retail price of normal butter during the sale period was 54 pence per 250 g, and the price of the sale butter in most supermarkets was 33–36 pence per 250 g. It is clear therefore that some of the EC subsidy was absorbed in the higher costs of handling butter withdrawn from store, and no doubt traders attempted to recover some of their losses through the disruption to normal trade the scheme caused. The total range of prices for Christmas butter extended from a minimum of 32 pence to a maximum of 42 pence per 250 g.

The analytical problem is how best to estimate the net improvement in butter sales resulting from the cheap sale. The commission of the EC made an estimate (without stating the method of analysis) that for the community as a whole 80 percent of the total allocation was substituted for butter that would have been purchased, and there was a net improvement to the market of 7 percent in the three months of the program. For the EC as a whole this amounted to a net increase in consumption of 39,850 tonnes out of the 200,000 tonnes. For the United Kingdom the EC commission estimated the substitution effect to be 74 percent and the increase of 10,000 tonnes to represent a 9 percent increase in the market.

At the Milk Marketing Board a univariate autoregressive Box-Jenkins analysis was used to predict the volume of sales of butter, packet margarine, soft margarine, and spreads from the AGB household panel data for the three months of the program and two months beyond that. Autoregressive methods of forecasting in the simplest terms regress the forecast variable on past values of the forecast variable itself (Box and Jenkins 1976, Chap. 5). Testing the accuracy of the forecasting models on past data against actual sales up to four periods ahead showed that errors for butter, packet margarine, and soft margarine never exceeded 4 percent. For spreads (i.e., mixtures of vegetable and dairy fats plus low vegetable fat products) the model performed less well, giving errors of up to 6 percent, but as this was the smallest sector of the market in tonnage terms, the error is tolerable.

The results of the comparison of the forecast sales to the actual sales from the panel are set out in Table 12.9. For the three 4-week periods ending March 23, 1985, the additional sales of butter would appear to have been some 13,000 tonnes, and there appears to have been a small carryover effect into the next two 4-week periods. Soft margarine sales lost some 6,200 tonnes in the campaign period and packet margarine some 4,000 tonnes, while spreads lost about 780 tonnes. This suggests that additional

Table 12.9. Comparison of forecast and actual household purchases of butter, margarine, and spreads in 1985 (4 weeks, ending . . .)

	January 26	February 23	March 23	April 20	May 18
			tonnes		
Butter					
Actual	18,885	17,297	15,605	12,757	12,621
Forecast	13,040	12,960	12,686	11,970	11,940
Difference	+5,845	+4,377	+2,925	+ 787	+ 681
Packet margarine					
Actual	4,882	4,594	4,730	4,786	5,157
Forecast	5,949	6,106	6,202	6,073	5,938
Difference	−1,067	−1,512	−1,472	−1,287	−781
Soft margarine					
Actual	16,946	17,624	17,697	18,311	18,350
Forecast	18,640	19,770	20,070	19,520	19,580
Difference	−1,694	−2,146	−2,373	−1,209	−1,230
Spreads					
Actual	3,413	3,343	3,461	3,621	3,951
Forecast	3,500	3,687	3,807	3,685	3,729
Difference	− 87	−344	−346	− 64	+222

Source: Milk Marketing Board, Economics Division.

butter purchases replaced margarine, but there might have been a slight increase in the combined sales of butter and margarine. Since butter spreads less well than margarine, consumers tend to use more of it than margarine when spreading it. However, this effect within these results is certainly within the levels of error of estimation in the model.

The model approach used by the Milk Marketing Board suggests a slightly larger net addition to total butter sales in the United Kingdom than the figures used by the EC commission. Moreover, the AGB household panel does not cover Northern Ireland, which probably received an allocation out of the 39,200 tonnes of some 2,000–3,000 tonnes. The EC Christmas butter program in 1984–85 was highly disruptive of general marketing, but the substitution effect was probably nearer two-thirds than three-quarters. It was nevertheless very expensive in terms of the net reduction in total stock achieved, the main reason that it is unlikely to be repeated.

CONCLUSION

No attempt has been made in this chapter to review all the work on demand analysis for dairy products in the United Kingdom. It is evident however that there are certain key uses for demand analysis.

1. Demand analysis is important to a general understanding of markets. Without it there is no understanding of why sales move in the way they do, and without such understanding intelligent marketing policies cannot be pursued.
2. Demand analysis can be used for forecasting both short term and long term. Short-term forecasts (crude or sophisticated) are a necessity for the immediate administration of business, that is, determining how much to produce for processors or paying producers on forward estimates for marketing organizations. Long-term forecasts are necessary for investment and strategic plans.
3. Demand analysis can assist in the making of decisions. Examples have been given of the level of generic advertising expenditure and brand pricing and brand advertising expenditure. This is a developing area of analysis and one that is expected to expand and improve in the future.
4. Demand analysis has an important use in the assessment of public policies when governments seek to intervene in the forces affecting supply and demand to achieve whatever goals believed desirable. The skills of the econometrician can be used to measure the efficiency with which the programs achieve their goals.

Finally, there is one area for the use of demand analysis for which no examples have been given in this chapter. Once demand analysis is able to develop realistic models to answer questions about generic and brand advertising, policy on relative prices, and other marketing issues, should it not then go a stage further and develop for large operators in the market a complete model of their activities to assist in formulating strategic plans? This would then enable the linking of decisions and demonstrate how they interact. This is perhaps ambitious, but it would seem to be one of the ways forward from the present state of thought, in which large and aggressive marketing organizations can put analytic and computing power together to improve their competitive positions.

REFERENCES

Box, G. E. P. and G. M. Jenkins, 1976. *Time Series Analysis: Forecasting and Control.* Holden-Day, San Francisco.

Brown, J. A. C. 1959. "Seasonality and Elasticity of the Demand for Food in Great Britain since De-rationing." *Journal of Agricultural Economics* 13(3):228–249.

Committee on Medical Aspects of Food Policy. 1984. *Diet and Cardiovascular Disease.* Report on Health and Social Subjects No. 28, Department of Health and Social Security, Her Majesty's Stationery Office, London.

Empson, J. D. 1958. "Economic Market Research and the Market for Milk." *Journal of Agricultural Economics* 13(2):169–182.

Federation of United Kingdom Milk Marketing Boards. 1987. *UK Dairy Facts and Figures.* Surrey, UK: Thames Ditton.

Milk Marketing Board. 1987. *EEC Dairy Facts and Figures.* Surrey, UK: Thames Ditton.

Milk Marketing Board. 1987. *Understanding the Butter Market—Ten Years On.* Surrey, UK: Thames Ditton.

MMD Ltd. 1986. A Quantitative Analysis of the Liquid Milk Market in England and Wales and the Impact of Generic Advertising. Report (unpublished) prepared for the National Dairy Council, London.

National Dairy Council. 1987. *Liquid Milk Report.* London.

National Food Survey Committee, 1974 to 1987. Ministry of Agriculture, Fisheries and Food. *Household Food Consumption and Expenditure.* Annual Reports. Her Majesty's Stationery Office, London.

National Economic Development Office, Food and Drink Manufacturing Economic Development Committee. 1987. *The Dairy Industry, Supply and Demand to 1990.* London.

Popkin, J. 1975. Some Avenues for the Improvement of Price Forecasts Generated by Macro-econometric Models. *American Journal of Agricultural Economics* 57:157–163.

Strak, J. 1983. Optimal Advertising Decision for Farmers and Food Processors. *Journal of Agricultural Economics* 34(3):303–315.

Strak, J. and L. Gill. 1983. *An Economic and Statistical Analysis of Advertising in the Market for Milk and Dairy Products in the UK.* Bulletin No. 189, Department of Agricultural Economics, University of Manchester.

Williams, R. E. 1986. "Forecasting Demand with Econometric Models." In *Workshop on Demand Analysis and Policy Evaluation*, edited by D. Peter Stonehouse. Bulletin No. 197/1986. International Dairy Federation, Brussels.

CHAPTER 13

Advertising Butter and Margarine in Canada

Ellen W. Goddard

GENERIC ADVERTISING of agricultural products is a marketing tool used increasingly in developed countries. Generic advertising is, in general, used as a method of increasing or maintaining market share or increasing producer revenues by causing a rightward shift of the demand function for a particular agricultural product. In Canada there are currently national generic advertising campaigns for butter, cheese, beef, pork, and eggs, and provincial advertising of milk.

The national butter advertising campaign (funded through producer levy) is of note because of the size of the advertising budget and because butter consumption has been steadily declining since the introduction of margarine into the Canadian market in 1949. This decline in per capita butter consumption is a serious problem for the dairy industry, due to the fact that 50 percent of industrial milk produced enters into butter production and this percentage represents 25 percent of all milk produced in the country. Prior to 1978 there was limited promotion of butter by various agencies but, in 1978, advertising of butter was formally taken over by the Dairy Bureau of Canada. Currently the Dairy Bureau's advertising budget (for butter, cheese, ice cream) ranks twenty-first among all advertising entities in Canada (McCarthy and Shapiro 1983, 556).

Apart from price competition from margarine, an additional factor leading to decreased butter consumption may be the perception by consumers of health problems related to cholesterol levels in butter. During

The author acknowledges the research assistance of Paul MacDonald and Apelu Tielu and the financial assistance of the Ontario Ministry of Agriculture and Food and the Ontario Milk Marketing Board.

the 1970s consumers became increasingly more health conscious and may have increased their substitution of "more healthy" margarines for butter (Dairy Bureau of Canada, personal communication).

Little public examination has been made of the effectiveness of the butter advertising campaign, or indeed, of the cross-commodity effects of the advertising of margarine and butter on the demand for each of these commodities. It is necessary to know the effects of the advertising campaigns for decisions on allocating funds for advertising.

LITERATURE REVIEW

With the increasing levels of generic advertising, studies analyzing the response of consumption (or sales) to advertising have been proliferating. One area that has been of great concern in the United States and United Kingdom is the effectiveness of milk advertising. The approach to analyzing the effectiveness of milk advertising has been to estimate single equations (of linear or another functional form) for the per capita consumption of milk with the price of milk and the price of a substitute drink, and with income and advertising as explanatory variables (e.g., Thompson and Eiler 1977; Kinnucan 1983). In these studies the hypothesis that advertising in a particular period has carryover effects in later periods was tested using different lag structures on the advertising variable. Both the studies mentioned found that advertising was a statistically significant variable in explaining milk consumption.

The single equation approach to estimating the demand for a particular product has deficiencies (although it is used consistently in the literature, largely due to data limitation reasons). For example, the cross-commodity effects of prices and of advertising are not clearly identified and it is not possible to test for restrictions suggested by consumer theory such as symmetry.

A few studies have concentrated on the consumption of butter and margarine and tried to relate these to advertising levels in the two markets. Pitts (1979) estimated separated regressions for the demand for butter and for margarine in Ireland as functions of prices of butter, margarine, and advertising of each commodity. He found little significance of the advertising variables. Strak and Gill (1983) examined the demand for a variety of dairy products in England. As part of this study, they estimated single equations to explain the demand for butter and margarine as functions of the prices of the two products, advertising variables, seasonal dummy variables, and lagged dependent variables. Their study showed no significant positive effect of generic advertising on butter sales.

From the literature, briefly surveyed, it appears that it has become accepted methodology to include advertising as an explanatory variable in a demand function for a particular good. More recently it has become apparent that other commodity advertising effects should also be included. However, the approach used has generally been single equation with few or no restrictions on cross-commodity results. The empirical literature surveyed also assumes that there is no differential response to different forms of advertising.

MODEL SPECIFICATION

In specifying a model that is appropriate for the examination of the impact of advertising on demand there are a couple of key decisions. The first decision relates to the general question of how advertising and consumer utility interact while the second decision relates to the definition of a manageable set of goods to examine given pragmatic data limitations.

Previous econometric time-series studies of the impact of advertising all included advertising as an explanatory variable in demand equations. Implicit in this specification is the inclusion of advertising variables (normally expenditure) as arguments in the underlying utility function. The inclusion of advertising in utility functions can be justified if advertising is interpreted as a taste change parameter operating directly on but not generating utility (Dixit and Norman 1979). A similar specification is used on the supply side when time is included directly in a profit or production function as a technology changing parameter (for a related discussion see Chapter 7). Some researchers argue that it is possible for advertising to contribute directly to utility [i.e., bandwagon effects (Fisher and McGowan 1979)] while others believe that the inclusion of advertising in a utility function is a "misrepresentation" of the role advertising plays in affecting consumer purchases (Kotowitz and Mathewson 1979). This research continues to use the inclusion of advertising in the utility function as a taste change parameter.

The empirical framework used to establish quantitative links between explanatory and dependent variables depends, for most researchers, on which set of restrictions they find easiest to live with. In estimating a single demand equation for a particular good the normal approach is to relate consumption of a good to a selected set of relevant prices (deflating both prices and income by a consumer price index). Consumers are assumed to respond to prices of all goods outside the selected ones by the same amount. Unless demands for more than one good are estimated it is not possible to establish whether consistent patterns of consumer behavior are being reported.

Alternately a two-stage demand system could be used as a method for simplifying the problems associated with modeling the demands for a group of goods without examining prices of all goods (Green 1971). A two-stage demand system rests on the assumption that consumers allocate their budgets across successively disaggregated bundles of goods (i.e., food, shelter, clothing), then disaggregate food expenditures into expenditures on fats, meats, and cereals, then disaggregate fats into butter and margarine. Economic theory suggests that the necessary condition for two- or multistage consumer optimization is that the marginal rates of substitution between goods (i.e., butter and margarine) be independent of any quantities of other goods consumed. This implies that if the demand for cereals (or individual cereal product) increases, the proportional consumption of individual fats would not change. The absolute level of individual fat consumption could change in response to a change in aggregate expenditure allocated to that branch of the decision tree. Figure 13.1 may prove helpful in clarifying the point.

The basic difference between a single equation approach and a two- or multistage demand system approach is the link between consumption at an individual good stage (e.g., butter) and disposable income. The restrictions associated with each of the approaches relate to the selection of goods or prices to be used in the analysis. In this research a two-stage demand system is used as the basic model for the analysis.

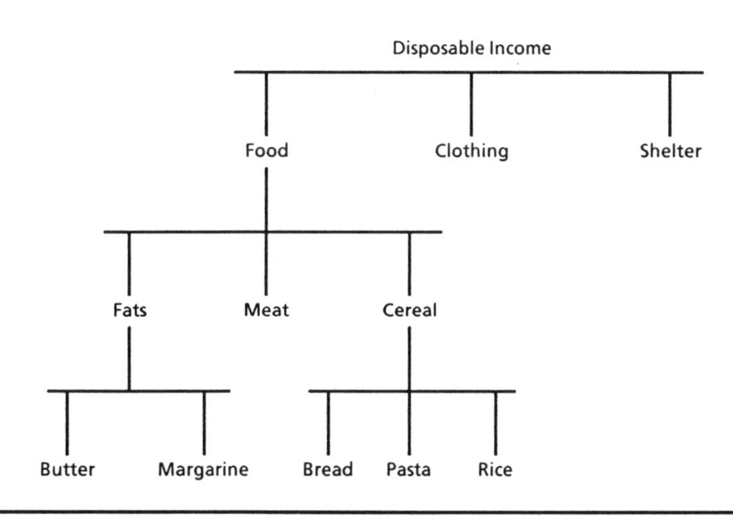

Figure 13.1. Decision tree

The first stage of the demand system proposed is a direct relationship between total expenditure on butter and margarine and traditional explanatory variables of price, income, and advertising expenditure on both goods.

$$TEXP_{BM} = f(WP_{BM}, A_{BM}, Y) \quad (13.1)$$

where WP_{BM} is the quantity weighted average price of butter and margarine, A_{BM} is the sum of advertising expenditures on butter and margarine, Y is disposable income, and $TEXP_{BM}$ is total expenditure on butter and margarine.

Seasonal dummy variables may be incorporated into the specification as may a lagged dependent variable to account for habits or persistence. Different functional forms may be tested with the above specification, for example, linear, double-log, or semi-log. The above specification (with appropriate deflation) is fairly traditional in empirical studies of individual goods. The problems of adding "apples and oranges" to generate an aggregate good are alleviated somewhat by the use of expenditure rather than quantity as dependent variable. A priori, one cannot sign the relationship between price and expenditure (as it depends on the elasticity of demand), but the signs on both income and advertising are expected to be positive for normal goods.

The second-stage equations are derived from a translog indirect utility function defined across normalized prices ($P_i/TEXP$) of butter and margarine and advertising expenditures associated with each good. The translog indirect utility function can provide a local second-order approximation to any arbitrary indirect utility function. This property is common to a set of functions that have come to be known as flexible functional forms. Other members of the set include the generalized Cobb-Douglas, the generalized Leontief, the generalized square root quadratic, and generalized Box-Cox. These functions are considered flexible in that they place few a priori restrictions on the full set of elasticities (price, income, substitution) at a base point (Caves and Christensen 1980).

The global properties of the different functional forms remain largely unknown except for empirical evidence from specific sets of data. In a variety of studies comparing functional forms and in explicit tests of the acceptability of certain functional forms (eg., Wales 1977; Appelbaum 1979; Goddard 1984; Amuah 1985) frequently the translog functional form could not be rejected. Using this criteria, the translog functional form was selected for this research.

A translog indirect utility function, in prices and total expenditure, is specified as

$$\ln V = \alpha_0 + \sum_{i=1}^{n} \alpha_i \ln P_i^* + 1/2 \sum_{i=1}^{n} \sum_{j=1}^{n} \beta_{ij} \ln P_i^* \ln P_j^* \quad i, j = 1, 2, \ldots, n \tag{13.2}$$

where $P_i^* = P_i / TEXP$ and $\beta_{ij} = \beta_{ji}$.

Using the logarithmic form of Roy's identity, expenditure shares for the ith commodity are

$$\frac{P_i X_i}{TEXP} = - \frac{\partial \ln V}{\partial \ln P_i} \Big/ \frac{\partial \ln V}{\partial \ln TEXP} \quad i = 1, 2, \ldots, n \tag{13.3}$$

For the translog indirect utility function (13.2) these are

$$\frac{P_i X_i}{TEXP} = W_i = \frac{\alpha_i + \sum_j \beta_{ij} \ln P_j^*}{\sum_i \alpha_i + \sum_i \sum_j \beta_{ij} \ln P_j^*} \quad i, j = 1, 2, \ldots, n \tag{13.4}$$

This basic translog model (Christensen and Manser 1977) does not include advertising or dynamic (lagged) effects. The addition of advertising expenditures on all goods in the group results in the indirect utility function

$$\begin{aligned} \ln V = \alpha_0 &+ \sum_i \alpha_i \ln P_i^* + \sum_j g_i \ln A_i + 1/2 \sum_i \sum_j \beta_{ij} \ln P_i^* \ln P_j^* \\ &+ 1/2 \sum_i \sum_j m_{ij} \ln A_i \ln A_j + \sum_i \sum_j c_{ij} \ln P_i^* \ln A_j \\ &\qquad i, j = 1, 2, \ldots, n \end{aligned} \tag{13.5}$$

Expenditure shares from (13.5) are

$$W_i = \frac{\alpha_i + \sum_j \beta_{ij} \ln P_j^* + \sum_j C_{ij} \ln A_j}{\sum_i \alpha_i + \sum_i \sum_j \beta_{ij} \ln P_j^* + \sum_i \sum_j C_{ij} \ln A_j} \quad i, j = 1, 2, \ldots, n \tag{13.6}$$

Simple habit formation models have been proposed to allow for dynamic effects on demand. Researchers test the hypothesis that there is an inertia to consumption behavior such that the level of consumption occurring in one period affects the consumer's response to changes in incomes and prices in another period. Manser (1976) and Blanciforti and Green (1983) have incorporated habit formation into their static models by incorporating dynamic elements into the intercept of their equations. For example, the α_i in the current specification can be assumed to depend linearly on consumption in the immediately preceding period.

$$\alpha_i = a_i + d_i X_{i_{t-1}} \quad i = 1, 2, \ldots, n \qquad (13.7)$$

Using (13.7), the expenditure share equation (13.6) can be written

$$W_i = \frac{a_i + d_i X_{i_{t-1}} + \sum_j \beta_{ij} \ln P_j^* + \sum_j C_{ij} \ln A_j}{\sum_i a_i + \sum_i d_i X_{i_{t-1}} + \sum_i \sum_j \beta_{ij} \ln P_j^* + \sum_i \sum_j C_{ij} \ln A_j} \quad i, j = 1, 2, \ldots, n \qquad (13.8)$$

A system of expenditure share equations must, by definition, sum to one. Thus, only $n - 1$ of the equations are independent. Since expenditure share equations are homogeneous of degree zero in prices and total expenditure, a normalization of parameters is necessary. The normalization used by Manser (1976) and in this study is that $\sum_i a_i = -1$. In estimation, additive error terms are assumed for the budget share equations. The error terms are assumed to have a joint normal distribution with mean zero and constant covariance. Woodland (1979) has addressed the problem of assuming that share equations have joint normal distributions and concludes that although the normal distribution may not be a theoretically appropriate specification, it may, for a large number of data sets, yield valid results.

THE DATA

Quarterly data for all explanatory variables were collected from 1973 to 1984. The data on sales of fats and oils were obtained from monthly data compiled from Statistics Canada (catalogue 32-006). These data comprised both household and industrial purchases of fats and oils and were the only disaggregated set of data available for this study. Data on population of Canada were obtained from Statistics Canada (catalogue 91-201). Data on prices, price indices, and the consumer expenditures index were obtained from Statistics Canada (catalogue 62-010). Data on per capita disposable income were obtained from Statistics Canada (catalogue 13-201).

The data for advertising expenditures were monthly for media advertising expenditures by commodity (Elliot Research and Media Measurement Corporation, Toronto). They represent expenditures for advertising on radio and television, which are the major promotional channels for fats and oils in Canada. For butter advertising expenditures, added information was obtained from the annual reports of the Canadian Dairy Commission.

The advertising figures for butter before 1978 represent the dollar expenditures by the Canadian Dairy Food Services Bureau (CDFSB) on the generic advertising of dairy products. After 1978 the expenditures on generic advertising of butter were made by the Dairy Bureau of Canada. Most butter advertising has been generic and there was no report of brand advertising expenditures for butter; therefore, only generic advertising expenditures on butter were considered.

The advertising expenditures for margarine include expenditures for the following brands: Monarch, Parkay, Imperial, Mom's, Blue Bonnet, Fleischmann's, Saffsweet, Thibault, Village, Blanchett, Gay Lea, Lactantia, Miracle, Butternut, Becel, and Achieve. Apart from these brands mentioned there was no report of any other brand advertising for margarine.

ESTIMATION PROCEDURE

Estimation of the aggregate demand function employed the ordinary least square (OLS) technique. Statistical inferences for the estimated aggregate demand equation are based on the t- and F statistics.

Estimation for the budget share equation (only one is estimated since they are not independent) was carried out by the maximum likelihood. Statistical inference was based on the likelihood ratio criterion. All computations were done with the computer package TSP (Time Series Processor), version 4.0B.

ANALYSIS OF RESULTS

To estimate the aggregate expenditure function and the expenditure share equation, the data described earlier were transformed into a per capita basis by dividing each series by Canada's population. Expenditures on each commodity were obtained by multiplying the per capita quantity data by deflated retail prices (deflated by the consumer expenditures index). Total expenditure on butter and margarine was obtained by adding consumer expenditure for the two commodities in this group.

For the aggregate expenditure function the price index was an endogenous quantity weighted average price for butter and margarine. Aggregate advertising expenditures were obtained by adding the per capita advertising expenditures on both commodities.

The results for the aggregate expenditure equation as given in Table 13.1 exhibit a strong positive response of total expenditure on butter and margarine to price (significant at the 1 percent level). This result is not surprising given the inelastic nature of demand for most food products. A priori, the expected response of total expenditure to changes in income levels was positive. The significant negative estimated response seems counterintuitive; but other published studies (Keane and Pitts 1985) have estimated negative expenditure elasticities for butter in some countries. It seems plausible that the general concerns about health may have had a negative impact on butter and margarine demand evidenced by a negative income elasticity.

Many different specifications of the total expenditure equation were tested with aggregate advertising as an explanatory variable. (It should be noted that pretesting on estimation of equations may inflate the value of the t-statistics. This is a more serious problem with the aggregate expenditure equation than with the budget share equation). Consistently the variable exhibited negative and insignificant coefficients and was dropped from the final specification. The hypothesis that advertising had a positive impact on aggregate demand for butter and margarine was rejected.

Table 13.1. Results from the aggregate expenditure model

Variable	Coefficient	t-statistic
Constant	1.709	3.12
Log weighted average price	.664	3.18
Log income	−.348	−2.28
Lagged dependent variable	.212	1.42
Quarter one dummy variable	−.123	−4.23
Quarter two dummy variable	−.108	−4.29
Quarter three dummy variable	−.052	−1.79
R^2	.64	
$\bar{R}^2$	.58	
D. W. Statistic	2.11	
h statistic	1.78	
F statistic	11.44	

Note: Dependent variable is the log of total expenditure on butter and margarine, sample period is 1973:3–1984:4.

The results in Table 13.1 show that habit formation or persistence buying on the part of consumers was a significant factor in determining expenditure on butter and margarine. The results also suggest significant seasonality in purchases of butter and margarine.

Different model specifications for the second stage were tested (i.e., prices; prices and advertising; prices, advertising, and habit persistence) with current variables. The same tests were conducted on a model using current prices and advertising expenditures lagged one period. More satisfactory results were found with the second specification.

The model was tested to establish the significance of each of the selected variables. The basic translog budget share system (basic translog model) includes only prices and total expenditure. To test for the significance of including advertising (lagged one quarter), the basic translog model was reestimated with advertising expenditures on butter and margarine also included as additional explanatory variables. Likelihood ratio test statistics were used to evaluate the restriction of not including advertising expenditure levels. Tests for the inclusion of habit formation were also similarly conducted. Again, the restricted model was the basic translog model. The final specification tested was obtained by combining price, advertising, and habit formation. The restricted model for this specification with price, habit formation, and advertising was the specification with price and advertising. The results from the testing procedure described here are presented in Table 13.2.

Table 13.2. Log of likelihood ratio test statistics for selection of model

		Log of likelihood function	Number of restrictions	Test statistics	Chi-Square 1%	Chi-Square 5%
Model 4	Prices, advertising, and habit formation (with symmetry)	106.06				
Model 3	Prices, advertising (with symmetry)	95.92	2	25.59	9.21	5.99
Model 2	Prices, habit formation (with symmetry)	103.59	3	4.94	13.28	9.49
Model 1	Prices (with symmetry)	89.53	(Model 4) 5	35.06	15.09	11.07
			(Model 3) 3	12.79	13.28	9.49
			(Model 2) 2	28.12	9.21	5.99

Note: Results are from estimation of the second-stage model for butter and margarine, sample period is 1973:3–1984:4.

From the results presented, restricted Model 1 can be rejected when compared to all other models (L.R. test statistics of 35.06, 12.79, and 28.12). The restriction of no advertising variables in Model 2 (prices and habit formation) compared to the unrestricted Model 4 (prices, habit formation, and advertising) cannot be rejected (L.R. test statistic of 4.94). This makes the significance of the advertising variables in the budget share equations unclear. The restrictions of not including either advertising or habit formation variables cannot be accepted (Model 4 compared to Model 1, L.R. test statistic of 35.06). Although statistically the advertising effects as a whole could not be found significant in all models, individual advertising coefficients were found to be significantly different from zero and, thus, the optimal model selected for the second stage of the demand was Model 4.

The above model specification tests were all conducted with symmetry ($\beta_{ij} = \beta_{ji}$) imposed on the price variables. While consumer theory suggests that symmetric price responses are rational the theory does not establish an equally strong case for symmetry of advertising responses. In this research it was decided to test for symmetry for both sets of variables. Tests were conducted on the short-term responses even in the model containing dynamic elements. Long-run responses are difficult to elucidate given the model estimated. The results from this testing with each of the four models are presented in Table 13.3. In three models (1, 2, 3) the symmetry restrictions on the price variables were statistically acceptable (in all cases at the 1 percent level and in some cases at the 5 percent level), and in Model 4 the symmetry restrictions on the advertising variables were acceptable. In Model 3 symmetry was rejected for both variables. As Model 4 has been selected as optimal, results were based on Model 4 with symmetry imposed on both variables.

Economic theory requires that an indirect utility function be monotonically decreasing in normalized prices ($P_i/TEXP$) and be strictly quasi-convex (Caves and Christensen 1980). The quasi-convexity requirement is equivalent to the requirement that the matrix of elasticities of substitution be negative semidefinite. Monotonicity requires that $\partial V/\partial P_i$ be strictly less than zero. This essentially means that the expenditure shares $W_1, W_2, \ldots, W_n$ be strictly positive (Caves and Christensen 1980).

Monotonicity requirements were met for each sample point but with the optimal model quasi-convexity requirements were not met for 20 percent of the sample points. In spite of this unsatisfactory result and because the model results were selected as the best out of a large number of alternatives on statistical grounds, it was decided to use the selected model as a base for estimated elasticities and simulations.

Results from the second stage of the model in terms of elasticities are presented in Table 13.4. As has been reported in previous studies the price elasticities show gross complementarity between butter and margarine in all

Table 13.3. Log of likelihood ratio test statistics for symmetry

		Log of likelihood function	Number of restrictions	Test statistics	Chi-Square 1%	Chi-Square 5%
Model 1	Prices, no advertising, no habit formation					
	Without symmetry	90.70				
	With symmetry	89.53	1	2.35	6.63	3.84
Model 2	Prices, with habit formation, no advertising					
	Without symmetry	105.57				
	With symmetry	103.59	1	3.98	6.63	3.84
Model 3	Prices, with advertising, no habit formation					
	Without symmetry on either	104.21				
	With symmetry on price	95.92	1	16.58	6.63	3.84
	With symmetry on advertising	100.49	1	7.44	6.63	3.84
	With symmetry on both	91.95	2	24.53	9.21	5.99
Model 4	Prices, with advertising, and habit formation					
	Without symmetry on either	108.14				
	With symmetry on price	106.06	1	4.17	6.63	3.84
	With symmetry on advertising	106.31	1	3.67	6.63	3.84
	With symmetry on both	104.74	2	6.80	9.21	5.99

Note: Results are for estimation of the second-stage model for butter and margarine, sample period is 1973:3–1984:4.

periods (Amuah 1985; Goddard and Amuah 1985). This result is robust across different sample periods and different functional forms tested for the model. On average when compared with previous results (Goddard and Amuah 1985) the butter own-price elasticities were larger than the previous estimates while the other price elasticities were smaller (in absolute value). The expenditure elasticities estimated in this study show a wider spread than those in the previous studies where they each approximated one. However, the results are consistent with previous estimates in that butter had the higher of the two expenditure elasticities.

From the substitution elasticities presented in Table 13.4 it is clear that the quasi-convexity requirements are not met for some sample points. By definition, the 1976:4, 1978:4 matrices are not negative semidefinite.

However, across the whole sample, for the observations where quasi-convexity requirements were met butter and margarine were shown to be net substitutes.

A priori, the own-advertising elasticities might be expected to have positive signs and the cross-advertising elasticities to have negative signs. These theoretically plausible results were not obtained for all sample points but were obtained for most, and held at the mean of the sample. Results from previous analyses had shown consistent negative responses to both butter and margarine advertising in the demand for butter. Thus, positive responses to advertising expenditure lagged one quarter seem more satisfactory with respect to a priori reasoning. Both the demands for butter and margarine appear to respond more to butter advertising than they do to margarine advertising.

Table 13.4. Butter and margarine price, expenditure, substitution, and advertising elasticities

	Price		Expenditure	Substitution		Advertising	
Period	Butter	Margarine		Butter	Margarine	Butter	Margarine
1974:4							
Butter	−0.80	−0.29	1.10	−0.25		0.008	−0.003
Margarine	−0.29	−0.56	0.86	0.37	−0.54	−0.012	0.004
1976:4							
Butter	−0.84	−0.55	1.39	0.04		0.02	0.002
Margarine	−0.27	−0.08	0.35	−0.07	0.13	−0.03	−0.003
1978:4							
Butter	−0.49	−0.58	1.07	0.23		0.012	−0.009
Margarine	−0.71	−0.19	0.90	−0.32	0.44	−0.017	0.013
1980:4							
Butter	−0.92	−0.34	1.26	−0.19		0.013	0.002
Margarine	−0.13	−0.41	0.54	0.33	−0.58	−0.023	−0.003
1982:4							
Butter	−0.86	−0.44	1.31	−0.07		0.016	0.001
Margarine	−0.23	−0.26	0.48	0.12	−0.20	−0.003	−0.002
1984:4							
Butter	−0.87	−0.46	1.33	−0.05		0.002	−0.002
Margarine	−0.22	−0.23	0.45	0.09	−0.16	−0.03	0.003
Mean							
Butter	−0.82	−0.36	1.17	−0.17		0.011	−0.001
Margarine	−0.26	−0.43	0.69	0.25	−0.37	−0.02	0.0012

STRUCTURAL SIMULATIONS

To investigate the dynamic nature of the optimal model the complete two-stage demand system was simulated with each of a single-period shock in the price of milk and in advertising expenditure on milk. The single-period shocks were a 10 percent increase in the level of each of the exogenous variables in the fourth quarter of 1973. The results for the particular endogenous variable are tabulated in Table 13.5. From the results in the table the exogenous single-period shocks produce results that decline without exception until they disappear.

SYSTEM SIMULATION RESULTS

The complete two-stage demand model was simulated under a variety of different scenarios. It is worthwhile pointing out that prices, disposable income, and advertising expenditure levels were all exogenous variables in the model with the endogenous variables being expenditure on butter and margarine, weighted average price of butter and margarine, expenditure shares and quantities demanded of both butter and margarine. This model specification rests on the assumption of perfectly elastic supplies of butter and margarine. In the long run, assumptions of perfectly elastic supplies of either are unrealistic; however, the assumption allows the simulation of the model in order to establish the upper bound of responses to shocks in the exogenous variables.

Table 13.5. Absolute differences in the quantity of butter consumed with 10 percent increases in the price of butter and in advertising of butter in 1973:4

		Quantity of butter consumed			
		Δ in the price of butter		Δ in advertising of butter	
Quarter	Base simulation	Total	Difference	Total	Difference
			Metric tons		
1973:4	35,961	34,606	−1,355	35,961	0
1974:1	31,491	31,451	−40	31,539	48
1974:2	29,026	28,930	−96	29,049	23
1974:3	29,507	29,439	−68	29,519	12
1974:4	32,335	32,293	−42	32,342	7
1975:1	29,698	29,680	−18	29,700	2
1975:2	28,324	28,315	−9	28,326	2
1975:3	28,705	28,699	−6	28,706	1
1975:4	31,253	31,249	−4	31,253	0
1976:1	28,151	28,148	−3	28,151	0
1976:2	26,562	26,561	−1		
1976:3	26,422	26,422	0		

Validation statistics for the base simulation of the model are provided in Table 13.6. The model captured major turning points in all endogenous variables and performed satisfactorily otherwise on statistical grounds.

The model was shocked by 10 percent sustained increases in the price of butter and in advertising expenditure on butter over the entire sample period. The results with respect to advertising expenditure were positive for the dairy industry (Table 13.7). For example, a 10 percent increase in advertising expenditure increased revenue net of advertising costs to the industry by .8 percent on average. This implied positive returns to increasing advertising expenditures by the Dairy Bureau of Canada. However, these returns were very small relative to the results of a similar study for fluid milk advertising in Ontario (Goddard and Tielu 1987).

It is possible that the Dairy Bureau is quickly approaching optimal advertising expenditure levels and increases in the budget should be carefully evaluated with respect to equivalent levels of return in alternate investments, whether these be in advertising of other dairy commodities or other forms of promotion. It should also be noted that these results are achieved without any impact on the aggregate expenditure (because the estimated advertising relationship was not significantly different from zero) on butter and margarine. The results are based solely on reallocation of expenditure between the two goods.

Increasing the price of butter also generated increased revenue to the industry. There was a significant increase in total expenditure on butter and margarine due to the inelastic nature of the aggregate expenditure equation. The decline in butter consumption was more than matched by the increase

Table 13.6. Validation statistics for two-stage butter and margarine demand model, base simulation

Selected variables	Mean	Correlation coefficient	Root mean squared error
TEXP (Butter and margarine) (per capita, deflated)	.58	.81	.03
WP_{BM} (Weighted average price)	.23	.95	.004
W_1 (Butter expenditure share)	.60	.75	.03
Quantity demanded butter (metric tons)	28,416	.75	2,118
Quantity demanded margarine (metric tons)	32,049	.81	2,534

Notes: Sample period is 1973:3–1984:4.

Table 13.7. Absolute and percentage changes in selected endogenous variables due to changes in the price of butter and in advertising expenditure on butter

		Advertising increased by 10%			Price increased by 10%		
	Base value	Actual	Difference	%	Actual	Difference	%
Total expenditure on butter and margarine (per capita)	0.6123	0.6137	0.0014	+0.2	0.6350	0.0227	+3.7
Quantity demanded butter	28,416.6	28,656.0	239.4	+0.8	26,796.9	1,619.7	−5.7
Quantity demanded margarine	32,049.6	31,782.3	267.3	−0.8	33,250.6	12.1	+3.7
Total revenue from sales of butter (Real $000)	43,151	43,521	370	+0.8	44,732	1,581	+3.7
Total butter revenue net of advertising costs (Real $000)	42,611	42,921	310	+0.8	44,192	1,581	+3.7
Advertising expenditure on butter (Real $000)	540	600			540		

Note: All values are at the mean of simulated values.

in price due to the inelasticity of demand for butter at the second stage of the model.

CONCLUSION

The research presented in this chapter attempts to determine the effectiveness of butter advertising in Canada.

On average own-advertising (lagged one quarter) elasticities were positive at the disaggregated (second-stage) level of the model. There was no significant relationship between advertising and aggregate expenditure on butter and margarine. Model simulations suggested that it would be possible for the Dairy Bureau of Canada to increase revenue net of advertising costs to the milk industry by increasing advertising expenditure. However, these increases were very slight, particularly when compared to similar results from studies for fluid milk. It is worthwhile to reiterate that the advertising effects were very small when tested in the model estimation stages and with different interpretation might reasonably have been constrained to zero. The industry might want to examine carefully the returns on a number of alternate investment possibilities before deciding to increase advertising expenditure on butter.

The results in this paper rest on a number of critical assumptions; for example, that advertising expenditures can be incorporated directly into utility functions as taste change parameters. One assumption that has not been explicitly highlighted is the rather limiting assumption that a dollar of advertising returns the same result regardless of which medium the advertising occurs in. With current time-series techniques and data availability it is difficult to relax this assumption. Future research efforts should be aimed at fine-tuning the analysis to provide the industry with more detailed information about different advertising strategies.

The empirical framework used in measuring advertising effectiveness in this research is one of a variety of different approaches that may be appropriate. Assumptions were made about the number of goods examined and the particular specification tested. Further testing might provide more evidence to substantiate the results reported in this chapter. However, it is possible that the results are sensitive to selection of goods and functional form.

The specification used to incorporate dynamic structure into the model was rudimentary and the examination of a more general approach (eg., Anderson and Blundell 1982) might prove fruitful. It might also prove fruitful to examine demographic factors and to try to establish the link between health concerns and sales of fats. A complete picture of the demand for goods is essential if strategies such as advertising are to be evaluated properly.

REFERENCES

Amuah, A. 1985. "Advertising Butter and Margarine in Canada." M.Sc. thesis, University of Guelph, Guelph.

Andersen, G. J. and R. W. Blundell. 1982. "Estimation and Hypothesis Testing in Dynamic Singular Equation Systems." *Econometrica* 50(6):1559–1571.

Appelbaum, E. 1979. "On the Choice of Functional Forms." *International Economic Review* 20 (2):449–457.

Blanciforti, L. and R. Green. 1983. "The Almost Ideal Demand System: A Comparison and Application to Food Groups." *Agricultural Economics Research* 35(3):1–10.

Caves, D. W. and L. R. Christensen. 1980. "Global Properties of Flexible Functional Forms." *American Economic Review* 70(3):422–432.

Christensen, L. R. and M. E. Manser. 1977. "Estimating U.S. Consumer Preferences for Meat with a Flexible Utility Function." *Journal of Econometrics* 10(2):728–729.

Dixit, A. and V. Norman. 1979. "Advertising and Welfare: Reply." *Bell Journal of Economics* 10(2):728–729.

Fisher, F. M. and J. J. McGowan. 1979. "Advertising and Welfare: Comment." *Bell Journal of Economics* 10(2):726–727.

Goddard, E. W. 1984. "Analysis of the International Beef Market." Ph.D. thesis, La Trobe University, Australia.

Goddard, E. W. and A. Amuah. 1985. "Advertising Butter and Margarine." In *Workshop on Demand Analyses and Policy Evaluation*, edited by D. Peter Stonehouse. International Dairy Federation Bulletin No. 197, Brussels, Belgium.

Goddard, E. W. and A. Tielu. 1987. *The OMMB's Fluid Milk Advertising*. Working paper WP87/14. Department of Agricultural Economics and Business, University of Guelph, Guelph.

Green, H. A. J. 1971. *Consumer Theory*. New York: Academic Press.

Keane, M. and E. Pitts. 1985. "Demand for Butter, Margarine and Cheese in the E.E.C. with Policy Implications." In *Workshop on Demand Analysis and Policy Evaluation*, edited by D. Peter Stonehouse. International Dairy Federation Bulletin No. 197, Brussels, Belgium.

Kinnucan, H. W. 1983. *Media Advertising Effects on Milk Demand: The Case of the Buffalo, New York Market*. Cornell University E.E. Res. 83-13.

Kotowitz, Y. and F. Mathewson. 1979. "Advertising, Consumer Information and Product Quality." *Bell Journal of Economics* 10(2):566–588.

McCarthy, E. J. and S. J. Shapiro. 1983. *Basic Marketing*. 3d cdn. ed. Homewood, Ill.: Richard D. Irwin.

Manser, M. E. 1976. "Elasticities of Demand for Food: An Analysis Using Non-Additive Utility Functions Allowing for Habit Formation." *Southern Economic Journal* 43(4):879–891.

Pitts, E. 1979. "Effects of Prices and Advertising on Demand for Butter and Margarine in the Republic of Ireland." *Irish Journal of Agricultural Economics and Rural Sociology* 7 (2):101–116.

Statistics Canada. *Consumer Prices and Price Indexes*. Catalogue No. 62-010, various issues.

_____. *Estimates of Population for Canada and the Provinces*. Catalogue No. 91-201, various issues.

_____. *National Income and Expenditure Accounts*. Catalogue No. 13-201, various issues.

_____. *Oils and Fats*. Catalogue No. 32-006, various issues.

Strak, J. and L. Gill. 1983. *An Economic and Statistical Analysis of Advertising in the Market for Milk and Dairy Products in the U.K.* University of Manchester Bulletin No. 189.

Thompson, S. R. and D. A. Eiler. 1977. "Determinants of Milk Advertising Effectiveness." *American Journal of Agricultural Economics* 59(2):323–330.

Wales, T. J. 1977. "On the Flexibility of Flexible Functional Forms." *Journal of Econometrics* 5:183–193.

Woodland, A. D. 1979. "Stochastic Specification and the Estimation of Share Equations." *Journal of Econometrics* 10:361–383.

CHAPTER 14

Advertising Fluid Milk in Ontario

Ellen W. Goddard, Henry W. Kinnucan, Apelu Tielu, and Evelyn Belleza

ADVERTISING FLUID milk in the electronic media is an important element of the Ontario Milk Marketing Board's (OMMB) market expansion program. A favorable trend in per capita milk sales throughout most of the 1970s seemed to validate the success of the advertising program. Real milk prices over the period were virtually constant. But in the late 1970s and early 1980s per capita fluid milk sales fell in Ontario at a time when annual advertising expenditures on fluid milk by the OMMB increased in nominal terms to over $4 million from about a quarter of a million in the early 1970s.

What do the apparently conflicting trends in sales and advertising imply about the efficacy of OMMB efforts to increase milk demand via media advertising? Has generic advertising of fluid milk in Ontario increased the demand for milk and, if so, is the increase large enough to compensate for the cost of the program? What impacts do OMMB price and advertising policies have on the demands for related beverages like soft drinks and fruit juices? A purpose of this chapter is to provide answers to these questions.

The problem of declining fluid milk sales in Ontario and the ability of industry-funded advertising efforts to offset this decline is important for several reasons. First, declining fluid milk sales adversely affect the profitability of dairying, even under supply management regulations. Ontario, with more than eight million people, is a major market for fluid milk. Second, more funds in nominal terms are being spent each year on generic advertising of milk and milk products. However, in real terms advertising budgets have remained relatively constant. Little is known about the efficiency and profitability of such expenditures.

Previous research examining commodity advertising effects has typically used one approach: a single equation demand model is estimated using time-series or pooled time-series and cross-sectional data in which advertising is incorporated in an ad hoc manner as a shift variable (e.g., Nerlove and Waugh 1961; Thompson and Eiler 1977; Kinnucan and Forker 1986; and Ward and McDonald 1986). Another approach in which a demand systems model is estimated using time-series data incorporating advertising also can be used to elucidate cross-commodity substitution and other effects of advertising (see Chapter 6). The results reported in this chapter differ from past research in that both approaches are applied to the same data set. In this way differences in estimated advertising effects due strictly to differences in modeling approach can be highlighted. An understanding and appreciation of such differences among researchers and policymakers in industry and government will lead to a more accurate assessment of the validity of econometric approaches to evaluating the impacts of advertising.

MEASURING THE IMPACT OF ADVERTISING

Model Specification

The empirical framework used to establish quantitative links between explanatory and dependent variables depends, for most researchers, on which set of restrictions they find most easy to live with. One approach is to relate consumption of a particular good to a selected set of relevant prices (deflating both prices and income by a consumer price index). Consumers are assumed to respond to prices of all goods other than selected ones through the consumer price index.

Alternately a two-stage demand system could be used as a method for simplifying the problems associated with modeling the demands for a group of goods without examining prices of all goods (Green 1971). A two-stage demand system rests on the assumption that consumers allocate their budgets across successively disaggregated bundles of goods (i.e., food, shelter, clothing, then disaggregate food expenditures into expenditures on vegetables, meats, and beverages, then disaggregate beverages into milk, juice, and soft drinks). The necessary condition for two- or multistage consumer optimization is that the marginal rates of substitution between goods (i.e., milk and juice) at one level of the decision process be independent of any quantities of other goods consumed.

According to Figure 14.1, this would imply that if consumption of any or all meats changed, expenditure allocated to beverages would adjust. However, this would only imply a change in the absolute level of individual

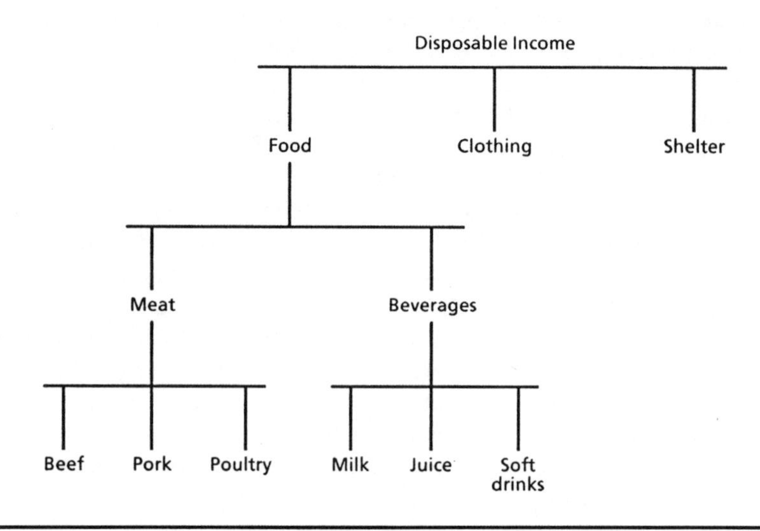

Figure 14.1. Multistage consumer decision tree

beverages consumed, not in the proportional share of each beverage in total expenditure on beverages.

The basic difference between a single equation approach and a two- or multistage demand system approach is the link between consumption at an individual good stage (e.g., milk) and disposable income. The restrictions associated with each of the approaches relate to the selection of goods or prices to be used in the analysis.

In this research both a two-stage demand system and a single equation model are used as the basic models for the analysis.

Data. The data used for both analyses were quarterly from the first quarter of 1971 to the fourth quarter of 1984. The commodities considered in the market for cold nonalcoholic beverages were fluid milk, soft drinks, tomato juice, apple juice, and orange juice. Disappearance and retail price data for Ontario were obtained from Statistics Canada as were data on disposable income, population, and the Consumer Price Index.

Advertising expenditures for each of the beverages were obtained from data compiled by Elliot Research and Media Measurement Corporation in Toronto. Media covered by these data include television, radio, and daily newspapers in Ontario. Advertising in consumer magazines specific to Ontario was estimated on a pro rata basis using expenditures at the national level. These estimates were added to data representing the other three media to obtain an aggregate measure of nonalcoholic beverage advertising in Ontario for each beverage. Expenditures for all commodities except fluid milk relate to brand advertising; advertising of fluid milk is generic and represents programs conducted by the Ontario Milk Marketing Board.

SINGLE EQUATION APPROACH

Following the tradition of earlier studies (Nerlove and Waugh 1961; Thompson and Eiler 1977; Kinnucan and Forker 1986), advertising effects on milk demand in the Ontario market were explored using a single equation model of the form

$$\ln QM_t = B_0 + B_1 \ln PM_t + B_2 \ln POJ_t + B_3 \ln INC_t + B_4 \ln AGE_t + B_5 \ln AD_t + B_6 \ln QM_{t-1} + \sum_{i=7}^{9} B_i S_{it} + B_{10} D_t + \epsilon_t \quad (14.1)$$

where $t = 1, 2, 3, \ldots, 56$ (1971 first quarter through 1984 fourth quarter); QM_t is Ontario fluid milk sales in liters per person per quarter; PM_t is Ontario retail price of milk in dollars per liter deflated by the Consumer Price Index, 1981 = 100; POJ_t is Ontario retail price of orange juice in dollars per liter by the CPI; INC_t is Ontario disposable income in dollars per person per year, deflated by the CPI; AGE_t is average age of Ontario population in years; AD_t is media advertising expenditures for fluid milk in Ontario in dollars per person per quarter, deflated by the CPI; S_{it} is seasonal dummy variables [i = 7 for winter (January, February, March), i = 8 for spring (April, May, June), and i = 9 for summer (July, August, September)]; D_t is the dummy variable to remove the effect of an outlier (1975.3 observation); and ϵ_t is a random error term.

A logarithmic specification was assumed to permit a diminishing marginal effect of advertising on sales (Simon and Arndt 1980). The lagged dependent variable was included to test the assumption that advertising and price carryover effects decline geometrically.

Expected signs of the parameters in equation (14.1) can be determined from theory and previous empirical research. Economic theory loosely suggests positive signs for B_2, B_3, B_5, and B_6 and a negative sign for B_1. Previous research (Kinnucan 1986) suggests an inverse relationship between age and milk sales; hence B_4 is expected to have a negative sign. Seasonality in milk demand is marked, usually declining in summer (Kinnucan and Forker 1986); hence, B_9 is expected to be negative. Depending on whether milk sales are higher, lower, or the same in winter and spring relative to fall, B_7 and B_8 may be positive, negative, or zero.

Regression results show equation (14.1) "explaining" 75 percent of the variation in milk sales (Table 14.1). The estimated parameters have anticipated signs in all cases except advertising. Income, age, seasonality, and dynamic effects are significant at the 1 percent level. The estimated advertising effect is not significantly different from zero. Elasticity values of −.095 for own-price, .065 for cross-price, and .348 for income agree with those obtained in other studies of milk demand (e.g., Kinnucan 1986 and 1987). The estimated parameter value of .348 for the lagged dependent variable indicates a stable and converging dynamic adjustment pattern. Significant seasonality in the demand for milk appears to exist in the Ontario market.

A caveat in interpreting the foregoing results is significant serial correlation in the residuals, evidenced by the large Durbin-h statistic. Correcting for serial correlation using the Prais-Winston procedure, however, resulted in sign reversals for several of the economic variables. Moreover, the estimated advertising effect remained insignificant in the corrected model.

Diagnostic checks indicated that the absence of a significant advertising effect could not be attributed to multicollinearity. However this and the presence of serial correlation suggest that the estimated model is highly sensitive to the specification and sample data. The robustness of results was investigated further by considering several variants of equation (14.1) including (1) redefining the dependent variable as (a) the market share of the total nonalcoholic beverage consumption and (b) total (rather than per capita) milk sales; (2) respecifying the advertising effect as (a) a finite distributed lag and (b) a moving weighted average of current and past aver-

Table 14.1. Ordinary least squares estimates of the fluid milk equation, Ontario

Independent variable	Coefficient		t-ratio
Intercept	2.735		2.88
Milk price	−0.095		−1.03
Orange price	0.065		1.29
Income	0.348		3.01
Age	−0.860		−2.48
Advertising	−0.001		−0.10
Lag dependent variable	0.342		3.55
S_7	−0.053		−3.91
S_8	−0.042		−3.74
S_9	−0.027		−2.51
D	−0.258		−8.46
R^2		.75	
$\bar{R}^2$		.70	
F		13.74	
h		3.86	

Note: Equation is estimated from 1971–1984.

age advertising; and (3) estimating alternative functional forms of equation (14.1). In each case the advertising effect remained insignificant at generally acceptable probability levels.

DEMAND SYSTEMS APPROACH

A demand systems approach to analyzing milk demand in the Ontario market was implemented by assuming a two-stage budgeting process (Green 1971). In the first stage the consumer decides what portion of total income to allocate to beverages and other (weakly separable) commodity groupings. In the second stage income allocated to beverages is distributed among the individual beverages within that grouping. Advertising is assumed to affect both stages of the budgeting process by altering consumers' tastes (Dixit and Norman 1979) both for broad commodity groupings and for individual consumer items. Utility maximization governs the allocation process in both stages. Specifications used to implement the systems approach are described below.

First-stage System. The first stage of the demand model was specified directly in logarithmic form as

$$\begin{aligned} \ln TEXP &= B_0 + B_1 \ln P_t + B_2 \ln Y_t + B_3 \ln A_t \\ &\quad + B_4 \ln TEXP_{t-1} + \upsilon_t \end{aligned} \tag{14.2}$$

where $t = 1, 2, 3, \ldots, 56$ (1971, first quarter through 1984, fourth quarter); $TEXP_t$ is total real expenditure on nonalcoholic beverages in Ontario; P_t is quantity weighted average price of the five beverages expressed in constant (1981) dollars; Y_t is per capita disposable income of Ontario residents expressed in constant (1981) dollars; A_t is aggregate per capita advertising expenditures for all five beverages expressed in constant (1981) dollars; and υ_t is a random error term.

The dependent variable in equation (14.2) was specified in expenditure rather than quantity form to permit consistent aggregation over commodities. From this it should be noted that the price coefficient will assume a positive sign if the demand for the cold nonalcoholic beverage groups is price inelastic. Other coefficients in the model are expected to have positive signs. As with the single equation model, a lagged dependent variable was specified in the first-stage model based on the hypothesis that habit formation can cause lags in consumers' response to changes in prices, income, and advertising.

Second-stage System. The second stage of the demand system is derived from a translog indirect utility function defined across normalized prices of the individual beverages, advertising, and demographics. Advertising and demographic variables (age) were incorporated into the basic translog model using the framework established by Jorgenson and Lau (1975). Following Manser (1976), advertising carryover effects and related dynamics were modeled by assuming that the intercept term of the demographic/advertising augmented utility function depends linearly on consumption in the immediately preceding period (Tielu 1987). A system of expenditure share equations expressing ordinary demands was obtained by applying the logarithmic form of Roy's identity to the indirect translog utility function modified to include the effects of demographics, advertising, and dynamics. The resulting second-stage system for estimating purposes was

$$w_i = \frac{a_i + r_i X_{i_{t-1}} + \sum_{j=1}^{5} B_{ij} \ln P_j^* + \sum_{j=1}^{5} C_{ij} \ln A_j + d_i \ln D}{\sum_{i=1}^{5} a_i + \sum_{i=1}^{5} r_i X_{i_{t-1}} + \sum_i \sum_j B_{ij} \ln P_j^* + \sum_i \sum_j C_{ij} \ln A_j + \sum_i d_i \ln D} \qquad i, j = 1, 2, \ldots, 5 \qquad (14.3)$$

where w_i is the expenditure share of the ith beverage; X_i is the quantity consumed of the ith beverage in liters per person per quarter; P_j^* is the normalized price of the jth beverage obtained by dividing the real price of the jth beverage by total expenditures on all beverages ($P_j/TEXP$); A_j is advertising expenditures for the jth beverage expressed in constant (1981) dollars per person; and D is the average age of the Ontario population.

Augmenting equation system (14.3) with additive error terms permits estimation of model parameters using maximum likelihood procedures. Since budget shares sum to 1, only four equations of the system are independent and need to be estimated; therefore, the equation for apple juice was dropped. Prior to estimation, the ratios of prices to total expenditures and advertising expenditure levels were scaled to equal 1 at the sample midpoint (1978.1). This rescaling has the virtue of simplifying estimation without affecting estimated elasticities or forecasted budget shares (Christensen and Manser 1977). Finally, because budget share equations are homogeneous of degree zero in prices and total expenditure, the $\sum_i a_i$ term in equation (14.3) was set to equal -1 following the normalization procedure suggested by Manser (1976).

The vector of additive disturbances, in each case, is assumed to be identically and independently joint normally distributed with mean vector zero and nonsingular covariance matrices. Woodland (1979) has addressed the problem of assuming that share equations have joint normal distributions and concludes that although the normal distribution may not be a theoretically appropriate specification it may, for a large number of data sets, yield valid results.

Results. OLS estimates of the aggregate (first-stage) model suggest that price, advertising, and habit formation are significant factors affecting consumption of nonalcoholic beverages in Ontario (Table 14.2). Included variables explain 84 percent of the total variation in consumer expenditures on these beverages. Income appears to be an unimportant demand shifter. The positive sign of the price coefficient is consistent with the hypothesis that the aggregate demand for nonalcoholic beverages in Ontario is price inelastic. The estimated aggregate long-run advertising elasticity of .078 is consistent with other studies that show advertising elasticities of mature consumer products tending not to exceed .10. The absence of serial correlation indicated by the Durbin-h statistic increases confidence in the validity of the first-stage specification.

Maximum likelihood estimates of the second-stage model (translog budget share system) was undertaken using a variety of tests to indicate the significance of included variables and the appropriateness of the theoretical restrictions relating to symmetry, monotonicity, and quasi-convexity (Goddard and Tielu 1988).

Results indicated that advertising, demographics (age), prices (expenditures), and habit formation were significant variables in the second-stage model. Symmetry was accepted for price effects but not for advertising. Budget shares were significantly positive, providing confirmation of the monotonicity restriction. The matrices of elasticities of substitution were

Table 14.2 Consumer demand for cold nonalcoholic beverages in Ontario

Variable	Coefficient		t-statistic	Total demand elasticities (short-run)
Constant	3.137		3.70	
Price	0.874		4.59	−.12
Income	−0.021		−0.20	.02
Advertising	0.059		3.34	.06
Lagged dependent variable	.229		2.23	
R^2		.85		
$\bar{R}^2$		.84		
Durbin-h statistic		.043		

Note: Ordinary least squares estimates of the first-stage (aggregate expenditure) model, 1971–84 quarterly data.

determined to be negative semidefinite for 80 percent of the sample, providing evidence that the underlying (indirect) utility function was quasiconvex. These results, taken together, suggest that the second-stage model was well specified in terms of theoretical consistency and inclusion of relevant explanatory variables, given the previous qualifications on the weak separability requirements of the two-stage system.

Uncompensated price elasticities and expenditure elasticities estimated from the second-stage model are presented in Table 14.3. Own-price elasticities were all negative in sign and their magnitudes indicate inelastic demands for each of the five beverages. The magnitude of the (short-run) own-price elasticity for fluid milk (−.293) was consistent with estimates from other studies (e.g., Kinnucan 1987).

The preponderance of negative signs for estimated cross-price elasticities suggests that Ontario consumers generally view these beverages as gross complements. Estimated expenditure elasticities ranged from .41 for apple juice to 2.26 for soft drinks. The expenditure elasticity for fluid milk was .43.

Advertising elasticities from the second-stage model are reported in Table 14.4. With the exception of orange juice, all own-advertising elasticities were positive. Cross elasticities of advertising are about evenly split between negative and positive signs, indicating an ambiguous effect of competitive advertising. For example, the estimated cross elasticities for fluid milk suggest that soft drinks, orange juice, and apple juice advertising increases the demand for milk but that tomato juice advertising decreases milk demand. Overall, apple juice advertising seems to have the smallest effects on demand (column) and milk advertising the largest. The demand for fluid milk seems least affected by advertising of other commodities (row) in absolute value terms.

Table 14.3. Price and expenditure elasticities for cold nonalcoholic beverages in Ontario

	Per 1% change in price of					
Percentage change in quantity of	Fluid milk	Soft drinks	Tomato juice	Orange juice	Apple juice	Expenditure elasticities
Fluid milk	−0.293	−0.068	−0.026	−0.018	−0.024	0.434
Soft drinks	−0.936	−0.802	−0.177	−0.002	−0.268	2.255
Tomato juice	−0.275	−0.032	−0.325	−0.099	−0.089	0.820
Orange juice	−0.554	0.241	−0.284	−0.852	0.142	1.307
Apple juice	−0.079	−0.096	−0.036	0.102	−0.302	0.410

Note: 1. Elasticities are evaluated at the sample midpoint (1978.4 observation).
2. Maximum likelihood estimates of the second-stage (translog budget share) model, 1971–84.

Table 14.4. Advertising elasticities for cold nonalcoholic beverages in Ontario

Percentage change in quantity of	Per 1% change in advertising of				
	Fluid milk	Soft drinks	Tomato juice	Orange juice	Apple juice
Fluid	.002	.004	−.002	.002	.002
Soft drinks	.001	.005	.005	−.031	−.005
Tomato juice	.070	.017	.006	.048	−.014
Orange juice	−.088	−.009	−.025	−.004	.036
Apple juice	−.040	−.028	−.005	.018	.003

Note: 1. Elasticities are evaluated at the sample midpoint (1978.4 observation).
2. Maximum likelihood estimates of the second-stage (translog budget shares) model, 1971–1984.

Model Simulation. To thoroughly investigate the properties of the two-stage demand model and to provide results of interest to policymakers, the complete model was simulated under a variety of scenarios. Of particular interest are responses to changes in advertising expenditure levels. However, for comparative purposes (and because the Ontario Milk Marketing Board has control over milk prices at the farm level) it is useful to examine the price change necessary to generate a similar change in industry revenue and that resulting from increasing advertising expenditure. Simulations were conducted in which the price of milk and milk advertising were each independently increased by specified amounts over the entire sample period. The simulations were all conducted assuming perfectly elastic supplies of fluid milk, a realistic assumption given the marketing system in Ontario.

Estimated impacts of these hypothetical actions by the OMMB on the cold nonalcoholic beverage market are indicated in Table 14.5. Ten percent increases and decreases in advertising expenditures generated approximately .2 percent changes in total revenue to the fluid milk industry. An equivalent change in revenue can be generated for the industry by increasing the real price of milk by .2 percent. The inelastic nature of demand for all beverages as well as that for individual beverages creates a situation where slight increases in price cause only negligible changes in quantity demanded.

Costs and returns to fluid milk advertising based on the foregoing simulations were computed using average values for farm price, advertising, and population. The results were generated assuming that supply was perfectly elastic under the Ontario marketing regulations implying that the quota would be relaxed by the full amount of the demand curve shift. From Figure 14.2 this is equivalent to shifting quota from Q_1 to Q_2. Under this scenario area *bcef* is the additional producer surplus produced by advertising. An alternate scenario might be that the quota remains fixed when advertising shifts the demand curve, market price rises to P^*, and *hj* is the extra producer surplus generated (accruing to the quota on production).

The data presented in Table 14.6 are based on a somewhat simpler scenario, as portrayed in Figure 14.3. The measure of returns to advertising above costs of production of $174,000 will be an overestimate of the area *bcef* in Figure 14.2 if there is a significant difference between MC_1 and MC_2. Using supply curve data generated in other research (Oxley 1988) it was found that the quantity increase was too small to result in a significant change in marginal costs. Over the range of data in this analysis the assumption of perfectly elastic supply and marginal cost curves is not unrealistic.

CONCLUSION

Results based on the demand systems model applied to Ontario data suggest that milk advertising significantly affects the demand for milk and related beverages. Advertising fluid milk not only increases the demand for fluid milk, it increases the demand for soft drinks and tomato juice and decreases the demand for orange juice and apple juice. Advertising conducted by the Ontario Milk Marketing Board appears to have increased milk demand sufficiently to offset the costs of the program.

Price and advertising, due to an inelastic demand for milk in Ontario, are complementary instruments with respect to OMMB marketing policies. That is, farm revenues can be enhanced by raising the level of either variable. For example, a 0.2 percent increase in the retail price of milk generates about the same increase in farm revenues as does a 10 percent increase in milk advertising. The issue of which instrument, price or advertising, is more efficient for accomplishing OMMB goals was not addressed in this research but offers an interesting problem for further study.

Compared to the demand systems model, the single equation model appeared to perform less favorably. Statistical significance of price effects is weak and the own-advertising effect is insignificant. Serial correlation is present and could not be purged without degrading the plausibility of parameter estimates. However, using the two-stage demand system model, in which statistically significant advertising effects were measured, required the use of restrictive assumptions such as weak separability between consumption of beverages examined in this research and all other goods. Models less restrictive than either of those tested in this study, in terms of consumer goods examined, might provide more insight into which set of restrictions is compatible with the real world.

Future research might examine a more general approach to incorporating dynamic structure into a demand system model (e.g., Anderson and

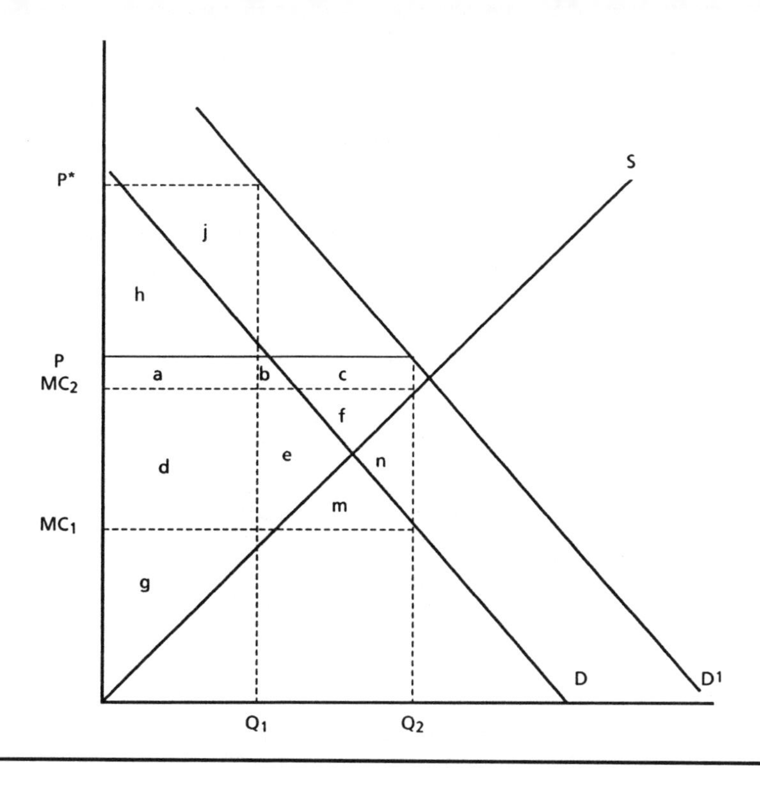

Figure 14.2. Impact of an advertising induced shift in the demand curve on the fluid milk market in Ontario

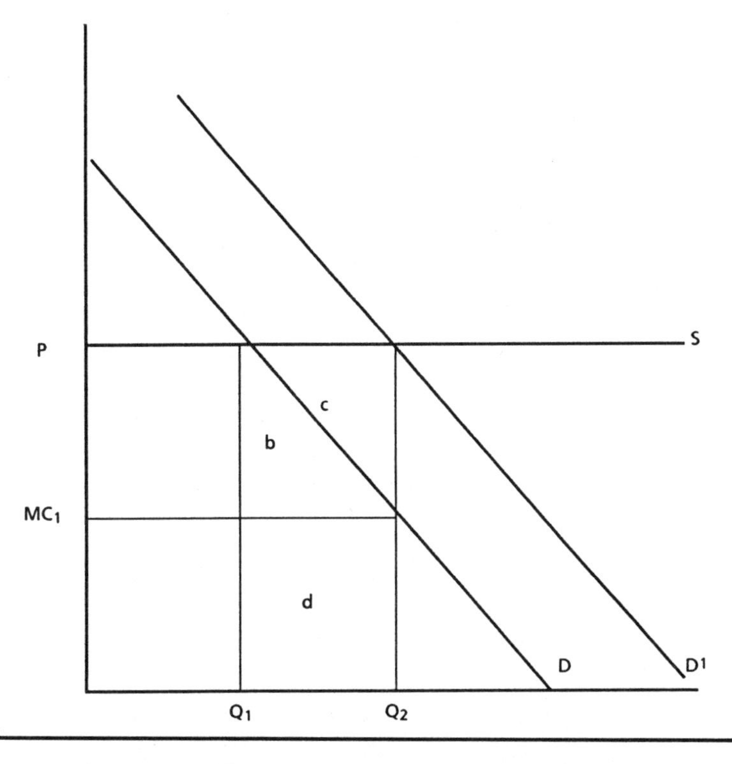

Figure 14.3. Impact of an advertising induced shift in the demand curve if supply is perfectly elastic

Table 14.5. Simulated impacts of milk price and milk advertising expenditure changes

	Base run	0.2 % Increase in PM	%Δ	10% Increase in AM	%Δ	10% Decrease in AM	%Δ	AM Constrained to zero	%Δ
TEXP ($ per capita)	49.358	49.408	0.10	49.430	0.14	49.285	−0.15	48.794	−1.14
QM (liters)	27.658	27.661	0.01	27.700	0.15	27.617	−0.15	27.297	−1.30
MTR ($000)	183,642	184,027	0.21	183,912	0.15	183,371	−0.15	181,442	−1.20
MNR ($000)	183,119	183,505	0.21	183,338	0.12	182,901	−0.12	181,442	−0.92
AM ($)	510,149	—	—	561,164	10	459,134	−10	1	−100
PM ($)	0.616	0.617	0.20	—	—	—	—	—	—

Note: Results are at mean of simulated values from two-stages demand model, 1971–84 sample period. (TEXP) total expenditure on cold nonalcoholic beverages; (QM) fluid milk demand; liters per capita; (MTR) total revenue for fluid milk; (MNR) net revenue for fluid milk; (AM) advertising expenditure on fluid milk; (PM) price of fluid milk.

Table 14.6. Estimated cost and returns to increased fluid milk advertising in Ontario

Data	Base with advertising		With 10% increase in advertising
Quantity of milk per capita per quarter (1/person)	27.658		27.700
Quantity of milk per capita per year (1/person)	110.6		110.8
Difference		0.2	
Retail revenue ($000)	734568		735648
Difference		1080	
Cost of increased advertising ($000)		204	
Retail revenue change less advertising cost ($000)		876	
Producer level return ($000) assuming an average marketing margin of $.359/liter		595	
Marginal cost of production ($000) (Oxley) $.29/liter		421 (area d Fig. 2)	
Returns to advertising above costs of production assuming elastic supply curve ($000)		174 (area b&c Fig. 2)	

Note: Calculation is at mean of simulation values.

Blundell 1982). Another area of fruitful research might be to examine a more general approach to the incorporation of demographic variables into demand systems (Lewbel 1985).

REFERENCES

Anderson, G. J. and R. W. Blundell. 1982. "Estimation and Hypothesis Testing in Dynamic Singular Equation Systems." *Econometrics* 50(6):1559–1571.

Christensen, L. R. and M. E. Manser. 1977. "Estimating U.S. Consumer Preference for Meat with a Flexible Utility Function." *Journal of Econometrics* 5:37–53.

Dixit, A. and V. Norman. 1979. "Advertising and Welfare: Reply." *Bell Journal of Economics* 10(2):728–729, autumn.

Goddard, E. W. 1988. *Modelling Advertising Effects in a Systems Framework*. Department of Agricultural Economics and Business WP88/1. Guelph: University of Guelph.

Goddard, E. W. and A. Tielu. 1988. "Assessing the Effectiveness of Fluid Milk Advertising in Ontario." *Canadian Journal of Agricultural Economics* 36:261–278.

Green, H. A. J. 1971. *Consumer Theory*. New York: Academic Press.

Jorgenson, D. W. and L. J. Lau. 1975. "The Structure of Consumer Preferences." *Annals of Economic and Social Measurement* 4(1):49–101.

Kinnucan, H. 1986. "Demographic Versus Media Advertising Effects on Milk Demand: The Case of the New York City Market." *Northeastern Journal of Agricultural and Resource Economics* 15:66–74.

_____.1987. "Effect of Canadian Advertising on Milk Demand: The Case of the Buffalo, New York Market." *Canadian Journal of Agricultural Economics* 24:181–196.

Kinnucan, H. and O. D. Forker. 1986. "Seasonality in the Consumer Response to Milk Advertising with Implications for Milk Promotion Policy." *American Journal of Agricultural Economics* 68:562–571.

Lewbel, A. 1985. "A Unified Approach to Incorporating Demographic or Other Effects into Demand Systems." *Review of Economic Studies* LII:1–18.

Manser, M. E. 1976. "Elasticities of Demand for Food: An Analysis Using Non-Additive Utility Functions Allowing for Habit Formation." *Southern Economic Journal* 43:879–891.

Nerlove, M. and F. V. Waugh. 1961. "Advertising Without Supply Control: Some Implications of a Study of the Advertising Oranges." *Journal of Farm Economics* 43:813–837.

Oxley, James. 1988. *An Ex Ante Analysis of Technological Change: Bovine Somatotropin and the Ontario Dairy Industry*. M.Sc. thesis, University of Guelph, Guelph.

Simon, J. L. and J. Arndt. 1980. "The Shape of the Advertising Response Function." *Journal of Advertising Research* 20:11–28.

Thompson, S. R. and D. A. Eiler. 1977. "Determinants of Milk Advertising Effectiveness." *American Journal of Agricultural Economics* 59:330–335.

Tielu, A. 1987. *A Quantitive Analysis of Advertising Fluid Milk in Ontario*. M.Sc. thesis, University of Guelph, Guelph.

Ward, R. W. and W. F. McDonald. 1986. "Effectiveness of Generic Milk Advertising: A Ten Region Study." *Agribusiness: An International Journal* 2:77–89.

Woodland, A. D. 1979. "Stochastic Specification and the Estimation of Share Equations." *Journal of Econometrics* 10:361–383.

CHAPTER 15

Demand for Manufacturing Milk

M. C. Hallberg

A FIRST PREREQUISITE to the development and implementation of sound policy for an industry is a thorough understanding of the nature of the economic relations and organizational structure characterizing that industry. The relevant economic relations can usually be quantified and, given the complexity of most industries, it is generally advantageous to do so. The organizational structure of an industry is less amenable to quantification. It is, nonetheless, imperative to understand this structure in order to formulate realistic quantitative relations and to aid one in interpreting and evaluating the resulting quantitative estimates.

A second prerequisite to the development and implementation of sound policy for an industry is to translate the above understanding into a form usable in the policy formation process. Economists have played an active role in the latter process (1) through the quantitative estimation of relevant economic and technical relationships, and (2) through the generation of "conditional forecasts" of outcomes of alternative policy choices using the relationships estimated. The term "conditional forecasts" is used here in the sense of Holt (1965) to mean forecasts for a specified time during the sample period or for a specified time beyond the sample period in which some factor (and most frequently a policy variable) takes on a value different from that actually observed (or that would be observed if policy remained the same). Hence conditional forecasts are distinct from what might be called "pure forecasts." The latter are projections into the future based on assumed values of *all* exogenous factors, including the relevant policy variables.

One important relationship (or set of relationships) required in this effort is(are) the demand function(s) for the product(s) of the industry. In the dairy industry demand functions for both fluid and manufactured products are required since consumers' responses to changes in such factors as prices or incomes can be expected to be quite different for these two product groups. Estimation issues include (1) the functional form of the equation(s) to be estimated, (2) the explanatory factors to be included in the equation(s), (3) the econometric technique to be used (for example, a single equation or simultaneous equations method), (4) the products to be aggregated for estimation and the aggregation procedure to be used, and (5) the level in the marketing chain at which the function(s) are to be estimated (retail, wholesale, or farm level).

In this chapter we deal exclusively with the demand for manufacturing milk and for manufactured dairy products. First, we examine the relative importance of manufacturing milk and of the products made from this milk and identify major trends in the manufacturing uses of milk. Second, we identify some of the salient conceptual issues for modeling the demand for manufacturing milk and manufactured dairy products. Finally, we review the results of past efforts at estimating demand functions for these commodities.

RELATIVE IMPORTANCE OF MANUFACTURING MILK AND RECENT TRENDS

The demand for manufacturing milk is, of course, derived from the retail demand of a wide variety of products, some consumed at retail, some used as inputs in the production of other foods, and some "consumed" by the government in its efforts to carry out the aims of the dairy price support program. Up to the 1960s somewhat more of the milk produced in the United States was used to produce fluid products than to produce manufactured products. Since the early 1960s, however, a majority of the milk produced has been used to produce manufactured dairy products. Indeed, over the last two decades manufactured product usage of the available milk supply in the United States has increased from about 48 percent to nearly 58 percent (Table 15.1). In part, this is due to a declining or stagnating demand for fluid milk products. In addition, high milk production levels in recent years brought about by high milk price supports, low feed prices, and/or technological advances have increased government purchases of manufactured dairy products—in particular, butter, cheese, and nonfat dry milk. Finally, it must be pointed out that over the past two decades the demand for manufactured dairy products in the aggregate has increased absolutely as well as relative to the demand for fluid products.

Table 15.1 also shows the general trends for the major products in the

Table 15.1. Milk production and utilization in the United States: 1966, 1976, and 1986

	1966			1976			1986		
	Product pounds	Milk equivalent	Percentage of total	Product pounds	Milk equivalent	Percentage of total	Product pounds	Milk equivalent	Percentage of total
SUPPLY									
Domestic production	—	119,892	98.9	—	120,269	99.8	—	144,080	99.8
Grade A[a]	—	82,725	68.3	—	97,418	80.8	—	125,350	86.8
Grade B[a]	—	37,167	30.6	—	22,851	19.0	—	18,730	13.0
Net imports and changes in storage[b]	—	1,371	1.1	—	258	0.2	—	270	0.2
Total supply	—	121,263	100.0		120,527	100.0	—	144,350	100.0
UTILIZATION									
Milk used in fluid products									
Milk sold to dealers	—	53,657	44.2	—	49,990	41.4	—	51,460	35.7
Milk sold direct to consumers	—	1,743	1.4	—	1,515	1.3	—	1,214	0.8
Total fluid use	—	55,400	45.6	—	51,505	42.7	—	52,674	36.5
Milk used in manufactured products									
Butter[a]	1,112.0	23,670	19.5	978.6	19,404	16.1	1,202.4	23,274	16.1
Cheese[d]									
American types	1,233.2	12,165	10.0	2,053.8	20,581	17.1	2,798.2	27,976	19.4
Swiss	136.7	915	0.7	196.3	1,190	1.0	227.3	1,264	0.9
Brick and Muenster	52.7	352	0.3	74.6	452	0.4	108.8	604	0.4
Cream and Neufchatel	116.4	1,112	0.9	168.0	1,455	1.2	321.5	2,553	1.8
Italian	271.1	1,813	1.5	747.4	4,531	3.7	1,632.8	9,077	6.3
Other	43.9	335	0.3	80.1	554	0.5	120.6	765	0.5
All cheese	1,854.0	16,692	13.8	3,320.2	28.763	23.9	5,209.2	42,239	29.3
Nonfat dry milk[d e]	1,579.8	—	—	926.2	—	—	1,293.8	—	—
Cottage cheese, creamed[d]	861.0	1,038	0.9	874.6	1,892	1.6	704.8	914	0.6
Canned milk[d]									
Evaporated	1,709.3	3,490	2.9	932.1	1,892	1.6	602.0	1,278	0.9

Table 15.1. *Continued*

	1966			1976			1986		
	Product pounds	Milk equivalent	Percentage of total	Product pounds	Milk equivalent	Percentage of total	Product pounds	Milk equivalent	Percentage of total
Condensed	487.2	1,149	1.0	270.6	643	0.5	299.3	835	0.6
Dry whole	94.4	698	0.6	78.1	554	0.5	122.4	901	0.6
All canned	2,290.9	5,337	4.4	1,280.8	3,089	2.6	1,023.7	3,014	2.1
Ice cream and frozen products	—	10,523	8.7	—	11,623	9.6	—	13,226	9.2
Other manufactured products	—	612	0.5	—	744	0.6	—	559	0.4
Total manufactured products	—	57,872	47.7	—	64,673	53.7	—	83,226	57.7
Milk used on farms where produced	—	5,481	4.5	—	2,966	2.5	—	2,559	1.8
Statistical discrepancy	—	2,510	2.1	—	1,383	1.1	—	−5,891	−4.1

Source: U.S. Department of Agriculture, *Milk Production, Disposition, and Income,* various issues; U.S. Department of Agriculture, *Dairy Products,* various issues.

[a] Estimated based on the percentage fluid grade milk reported in U.S. Department of Agriculture, *Milk Production, Disposition, and Income.*

[b] Net imports of ingredients and net changes in storage of cream.

[c] Creamery butter production less the milk equivalent of butter produced from whey cream.

[d] Product pounds as reported in U.S. Department of Agriculture, *Dairy Products.* The milk equivalent of the different types of cheese was estimated on the basis of the average butterfat content of the cheese type and then adjusted so the sum equaled the reported total.

[e] Nonfat dry milk is assumed to be a by-product of other manufactured products so the milk equivalent of nonfat dry milk is zero.

manufactured products sector. Butter production has declined slightly while canned milk production has declined nearly 100 percent. The big increase has been in cheese production. In total, cheese production as a percentage of total availability of milk has increased from 14 percent in 1966 to 29 percent in 1986. American cheese production has more than doubled over this period while Italian cheese production has increased fivefold from 1.5 percent of available supply of milk in 1966 to 6.3 percent of available supply of milk in 1986.

A variety of factors might explain these trends in cheese production and utilization. There has undoubtedly been some substitution of cheese for meat in the typical American diet. Cheese has also become a popular snack-food before and after meals. Further, some have noted a strong positive correlation between cheese consumption and wine consumption in recent years. The phenomenal growth in use of Italian cheese is most likely attributable to the growth in popularity of pizza. Increases in both American and Italian cheese consumption have probably also been influenced by the increase in away-from-home eating habits of American consumers and, in particular, by the increase in eating at fast-food restaurants.

Clearly, the demand for manufacturing milk is much too important to treat lightly, and estimation of the demand function for manufactured products and for manufacturing milk can be expected to be complicated not only by the fact that a variety of final products must be considered but also by the fact that the relative importance of the different products in the total of manufactured dairy products has been changing in recent years.

CONCEPTUAL ISSUES

The Market for Manufacturing Milk

There are, in effect, two "markets" for manufacturing milk: a competitive Grade B market and a residual or surplus Grade A market. The price structure in both of these markets is buttressed by the dairy price support program. This program is designed to support the average price farmers receive for their manufacturing milk at a level determined by the Congress or secretary of the U.S. Department of Agriculture, within the rules specified by Congress. Historically this support level has been set at a percentage of the parity price for manufacturing milk, although currently other rules are invoked. The support level is intended to provide a floor below which the market price of manufacturing milk will not fall. The program is maintained primarily through a system of open market purchases and sales by the Commodity Credit Corporation (CCC) of butter, cheese, and nonfat dry milk manufactured from raw milk (see Hallberg and King

1980).

Grade A milk is milk that meets U.S. Public Health Service sanitary standards for drinking quality milk. As can be seen from Table 15.1, nearly 87 percent of the milk currently produced is Grade A milk, although this was not always the case. Grade B milk need not meet these high sanitary standards and, consequently, cannot be used in the production of fluid products but must be used in the production of such products as butter, nonfat dry milk, and cheddar cheese. Because Grade B milk need not meet such high sanitary standards and should therefore be less costly to produce, it does not command as high a price as does Grade A milk.

The actual pay price for Grade B milk is determined in a freely operating, competitive arena usually associated with an area of the greatest concentration of Grade B milk production, Minnesota and Wisconsin (see Hallberg and King 1980). This is a price that presumably reflects all the relevant forces affecting supply and demand. The (minimum) price that handlers regulated by federal milk marketing orders must pay producers for "surplus" Grade A milk (milk in excess of that needed to satisfy fluid milk needs), on the other hand, is set by the federal milk marketing order to which this milk is shipped. The price for surplus Grade A is set by formula and, in fact, is always at or near the average competitive pay price for Grade B milk recorded in the Minnesota-Wisconsin area.

Farmers delivering Grade A milk to a handler regulated by a federal milk marketing order receive a single price for their milk. This price is a weighted average or blend of the prices of (1) Grade A milk used to produce fluid products and (2) surplus Grade A milk. Both of these prices are determined by formula established by the federal milk marketing orders as is the weighting (or pooling) procedure to be used.

In recent years there has been a substantial decline in the production of Grade B milk (see Table 15.1) as farmers have converted from Grade B to Grade A production. The blend pricing system of the federal milk marketing orders has provided the incentive for this to occur. That is, the difference between the blend price farmers get for milk delivered to the federal milk marketing order and the price they would get on the open market for Grade B milk has generally been higher than the Grade A–Grade B production cost differential for most producers. Hence, the price differential has favored conversion and rational farmers have acted accordingly. The result, however, has been to drive the blend price down to the point where few if any of the remaining Grade B producers can afford to convert.

This brief review of government intervention in the U.S. dairy industry provides the background for developing a simple graphical model to depict the essence of price determination in the industry. In Figure 15.1 the line labeled D_fD_f represents the commercial demand for fluid milk and the line labeled D_mD_m represents the commercial demand for manufacturing milk. The line labeled *SS* represents the supply of all milk. (There are in fact two supply curves—one for Grade A milk and one for Grade B milk—and the price a farmer receives for milk will differ depending upon which grade of milk is produced.) Line $D_f(L)AR$ represents the average price (revenue) farmers receive for milk, the blend price due to the classifying, pricing, and pooling procedures of the federal milk marketing orders. Line KD_g indicates that at price P_s demand for manufacturing milk is infinitely elastic because the CCC is authorized to purchase all of the butter, cheese, and nonfat dry milk that is produced at that raw milk price, preventing the price of manufacturing milk from going below P_s.

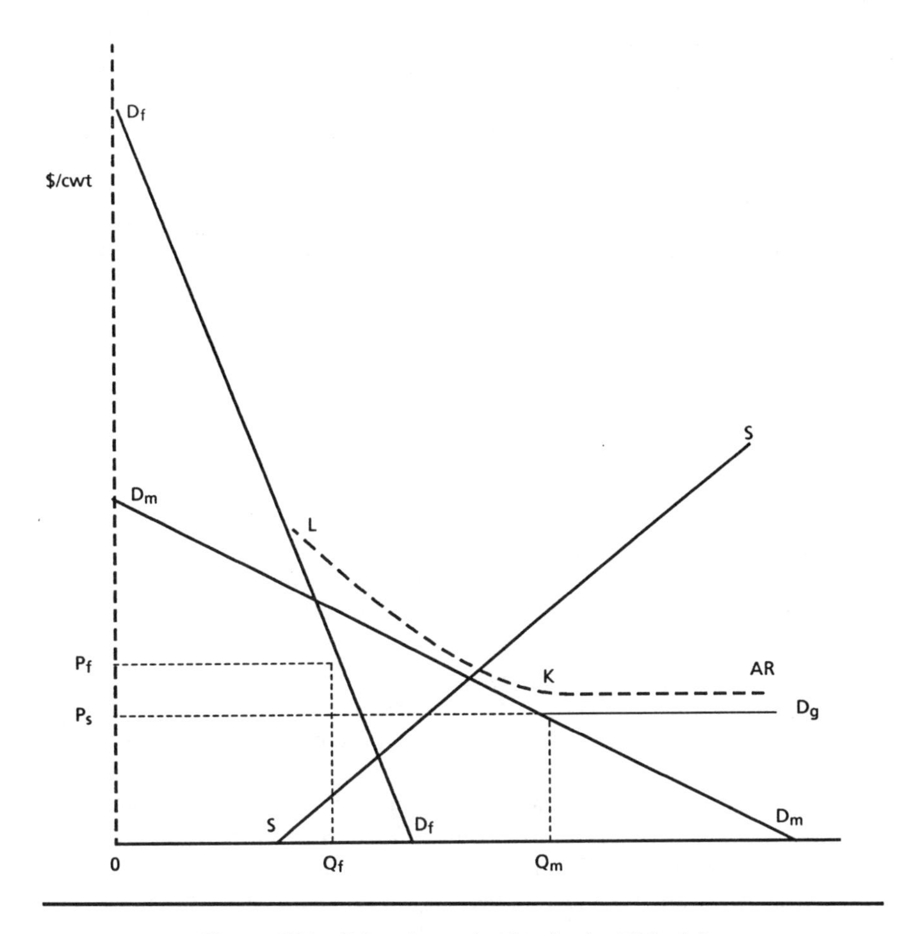

Figure 15.1. Price determination in the U.S. dairy sector with the dairy price support program

Some writers on the subject of milk pricing draw the demand curve for manufacturing milk perfectly horizontal at P_s beginning at the vertical axis (see, for example, Kessel 1967). This is misleading, however, because even with a price support program in place, market price will be above the support price during periods of short supply. By the same token, it is erroneous to ignore the horizontal portion of the manufacturing milk demand curve as if the government policy in effect did not exist.

Clearly a kink exists in the demand function of manufacturing milk as long as the dairy price support program is in effect. The position of the kink, however, obviously changes every time a new P_s is established by the U.S. Congress or secretary of agriculture.

Figure 15.1 is an admitted oversimplification of the determination of milk prices in the U.S. dairy industry. One should distinguish between Grade A and Grade B milk. Further, one should explicitly account for the individual products (both fluid and manufactured) produced by the industry. Finally, the determination of "make allowances" (see Manchester 1983), which affect announced CCC purchase prices for butter, cheese, and nonfat dry milk, should be reflected. All this, however, would make the system unmanageable graphically. Furthermore, the ideas expressed by Figure 15.1 are sufficient to illustrate a very important concept—namely, that the demand for manufacturing milk is not only more elastic in general than is the demand for fluid milk, but it also becomes infinitely elastic at the government price support level. Since CCC purchase prices for butter, cheese, and nonfat dry milk are derived from the support price for manufacturing milk, we can expect the demand functions for these products also to be kinked at the announced purchase prices.

Demand Estimation

The demand function for manufacturing milk can be estimated directly using per capita commercial disappearance of manufacturing milk as the dependent variable and the market price of manufacturing milk as one of the independent variables. Other independent variables may include per capita disposable income, some variable or variables to represent the changing age distribution of the population, a variable representing the increased importance of away-from-home eating, and perhaps a time trend to reflect a general change in tastes over time.

The demand function for manufacturing milk may also be derived from the demand functions for products manufactured from this milk, each of which was estimated, for example, at the retail level with a similar set of independent variables. Conceptually this approach is quite straightforward. One would simply transform each such demand function from the product form at retail to the raw milk equivalent and aggregate. This would involve

(1) multiplying each product demand function by a factor representing the yield of that product from raw milk and (2) converting the term involving retail price of the product so that it is now a function of farm level price (using an estimated price transmission equation) instead of the retail-level price. The equations so transformed could then be aggregated to obtain the demand function for manufacturing milk. This procedure is clearly appropriate if and only if the demand functions for the included products are independent and if each commodity can be viewed as a variety of the basic commodity, raw milk.

Practically, however, it is not so straightforward to derive a demand function for manufacturing milk in this latter manner. First, data for all the manufactured products are not readily available or of equal reliability, and some of the minor products must inevitably be excluded or estimated with lower precision. Second, while the demand functions themselves may be independent, the production processes for the products are not necessarily so. That is, some manufactured products are produced as joint products in fixed proportions. Examples include butter and nonfat dry milk, butter and condensed skim milk, cottage cheese and cream, and cheddar cheese and whey. This means that estimating the yield of products from raw milk with which to transform the product demand functions will be subject to error. Further, estimating the price transmission equations is not always easy or precise.

To analyze the full range of policy issues, however, it is usually necessary to include the demand functions at retail (or at wholesale) for the major products in order to preserve detail on individual commodities. Here it will generally be sufficient to estimate demand functions for butter, cheese, nonfat dry milk, and perhaps evaporated and condensed milk and frozen products. Retail-to-farm level price transmission equations will then be used to close the model.

EMPIRICAL ESTIMATES

A variety of econometric techniques has been used in demand analysis for dairy products, ranging from ordinary least squares (OLS) to estimation within the context of complete demand systems. The latter can encompass a wide spectrum of commodities and attempts to capture empirically the effects of commodity interdependencies (i.e., substitution and/or complementarily effects) on the estimated parameters. OLS is capable of capturing some of these interdependencies but multicollinearity problems generally prevent the examination of any more than one or two such factors.

Use of OLS does permit the examination of a variety of exogenous factors (such as the changing age distribution of the population and changing tastes). The complete demand system approaches should also permit an incorporation of such factors but typically do not. This remains a major limitation of all complete system approaches except that of Heien and Wessells (1988). Much of the past work in demand analysis has shown that these exogenous factors for certain commodities, and particularly for dairy products, are more important determinants of demand than are the more traditional price and income variables. The aging of the population, for example, most assuredly has had a significant impact on the declining consumption of fluid milk and on the increased consumption of some of the manufactured products. Consumer concern about cholesterol intake has also been an important factor over the past 10 years or so (see Warland et al. 1989). Finally, U.S. consumers' increased tendency to consume meals away from home (which may or may not be income related) has impacted dairy product consumption.

As the summary in Table 15.2 shows, several researchers have estimated demand functions for the more important manufactured dairy products with varied results. Only three studies of recent vintage have attempted to estimate the demand for manufacturing milk directly. In this table we distinguish between short-run and long-run elasticity estimates for each commodity. This is somewhat hazardous because of the difficulty in specifying the appropriate length of run. In general, if the authors of the studies cited did not specify the length of run or if the model used did not incorporate an adjustment mechanism with which to capture long-run effects, we classified the result as a short-run estimate. With the exception of the studies by Boehm and Babb (1975) and by Thraen et al. (1983), all of the estimates shown in Table 15.2 were based on annual observations.

Rojko (1957) was one of the first economists to estimate demand functions for manufactured dairy products. His work was based on the most sophisticated technique available at the time—ordinary least squares (OLS)—and did not take account of the full range of interdependencies in the food demand system. Subsequent work attempted to rectify this shortcoming through the use of complete demand system estimation techniques. The later work is illustrated by that of Brandow (1961), George and King (1971), Heien (1982), Huang (1985), and Heien and Wessells (1988). The results obtained by these authors were quite mixed.

Rojko estimated the demand elasticity of butter to be −1.37. Subsequent studies suggest it is much lower. Huang's work, for example, which represents a quite recent demand study and is the most sophisticated, yields a butter elasticity of −0.167. The declining importance of butter in the American consumer's food budget coupled with increased family incomes as compared with the immediate post–World War II period probably supports an elasticity somewhat lower than −1.37. On the other hand, an elasticity as low as that estimated by Huang is, for most people knowledgeable about this industry, probably equally unacceptable.

Rojko's estimate of the elasticity of cheese appears to have stood the test of time somewhat better. Again, Huang's estimate for cheese is quite low, but perhaps not significantly lower than that of George and King. Part of the difficulty here may well be that "cheese" is too broad a category, at least for recent years. In this connection it is interesting to note that in a recent study by the USDA (1987), the elasticity of demand for natural cheese was estimated to be −.737. These results were achieved using cross-sectional data over the period January 1982 to June 1986.

It should be noted that in almost every study dealing with cheese, the researcher records some considerable difficulty in obtaining a reliable estimate of the price elasticity (indeed in most cases in even obtaining a negative coefficient). One might suspect a principal reason is the fact that there has been such a phenomenal growth in cheese demand since the mid-1950s that this growth effect has overshadowed the price effect. Thus, until we can successfully model this growth phenomenon, we shall undoubtedly continue to have difficulty separating out the price response for cheese.

As Table 15.2 indicates there is also much variation in the elasticities estimated for the remaining manufactured dairy products. The study by Huang (1985) generally included the most extreme estimates. It may well be that the results of Huang's study are the most accurate. It is, for example, difficult to anticipate a reasonable value for the elasticity of demand for canned milk. However, the demand for ice cream and related frozen products is likely not so nearly perfectly inelastic as suggested by the Huang study.

The three studies reporting the elasticity of demand for manufacturing milk yield more consistent estimates of the price elasticity. This is not surprising in view of the fact that the estimating models and techniques as well as the sample periods used were similar. Clearly, the demand for manufacturing milk in the aggregate appears to be responsive to changes in the price of raw milk and to be somewhat more responsive to price than is the demand for fluid milk.

CONCLUSION

Milk used to produce manufactured dairy products now exceeds that used to produce fluid products. For the most part this can be attributed to the increased commercial disappearance of manufactured products and, in particular, cheese. In fact, it might be said that the increased consumption of cheese in the last decade constitutes the one bright spot for the U.S. dairy industry over the period. Hence, it is just as important, if not more so, that we seek to increase our understanding of the structure of demand for

Table 15.2. Estimated retail demand elasticities for manufactured dairy products and for farm-level demand elasticity for manufacturing milk (short-run and long-run)[c]

	Sample period	Estimation method	Butter		Cheese		Evap/Cond		Frozen		Nonfat dry		Manufacturing Milk	
			S.R.	L.R.	S.R.	L.R.	S.R.	L.R.	S.R.	L.R.	S.R.	L.R.	S.R.	L.R.
Rojko	1947–54	OLS	−1.37	—	−0.75	—	—	—	—	—	—	—	—	—
Brandow	1955–57	System	−0.85	—	−0.70	—	−0.30	—	−0.55	—	—	—	—	—
George/King	1962–66	System	−0.65	—	−0.46	—	−0.32	—	−0.53	—	—	—	—	—
Boehm/Babb[a]	1972–74	OLS	−0.73	−0.76	−0.85	−0.85	—	—	−0.42	−0.69	−0.45	−2.24	—	—
Huang	1953–83	System	−0.17	—	−0.33	—	−0.83	—	−0.12	—	—	—	—	—
Hallberg/Fallert	1955–73	OLS	(−0.7)[b]	—	(−0.5)	—	−0.6	—	−0.33	—	(−0.5)	—	—	—
Thraen et al.	1972–73	Tobit	−0.22	−0.70	−0.13	−0.62	−0.07	−0.25	−0.11	−0.23	−0.42	−1.92	—	—
Heien	1947–79	System	−1.93	—	—	—	—	—	—	—	—	—	—	—
Thraen/Hammond	1949–78	OLS	—	—	—	—	—	—	—	—	—	—	−0.20	—
Dahlgran	1968–77	OLS	—	—	—	—	—	—	—	—	—	—	−0.35	—
Hallberg	1955–78	OLS	—	—	—	—	—	—	—	—	—	—	−0.16[d]	−0.43[d]

[a] Estimated from panel data.

[b] Coefficients within parentheses here were assumed a priori and the remaining demand function coefficients were estimated by OLS subject to this restriction.

[c] Short-run elasticities here represent the response to price changes to consuming households actually purchasing the product. Long-run elasticities include the adjustment due to households' entry into or exit from the commodity market.

[d] Using the identical model and data through 1985, these elasticities are estimated to be −0.18 and −0.66, respectively.

the manufactured dairy products sector as it is to seek to increase our understanding of the demand for fluid dairy products. Accurate knowledge of the demand relations for manufactured dairy products is also critical, of course, to the establishment of rational short-run as well as long-run price policy for the entire dairy industry.

Several researchers have estimated demand relations for various manufactured dairy products. A variety of models, estimation techniques, and sample periods have been used. The results have been mixed and in some cases controversial. Two conclusions seem to emerge from the work to date. First, the demand for manufacturing milk derived from the demand for the products produced from this milk appears to be more price elastic than that of milk used to produce fluid products, but it is still quite price inelastic at least in the short run. Second, the diversity among the estimated elasticities suggests there is room for further and innovative work in this area.

A review of the empirical work to date in estimating the demand functions for manufactured dairy products suggests that care must be exercised to include as explanatory variables all relevant nonprice and nonincome variables; for example, the changing age distribution of the population, changes in tastes and preferences, substitute and complementary products, and so on. The available evidence seems to point to the fact that such variables may be more important in explaining recent variations in demand than prices and income. This is a fairly logical conclusion given the relatively small variation in real prices and incomes experienced in recent years.

Finally, the functional forms underlying the models estimated to date should be reexamined, at least for the major manufactured products like butter, cheese, and nonfat dry milk. This brief review of the theoretical underpinnings of this market has revealed that the demand for manufacturing milk, and hence for the products produced from this milk, is inherently nonlinear. It may well be that estimation procedures dealing explicitly with discontinuous or kinked functions need to be used. At minimum, models linear in prices must be scrutinized carefully.

REFERENCES

Boehm, William T. and Emerson M. Babb. 1975. "Household Consumption of Perishable Manufactured Products: Frozen Desserts and Specialty Products." Purdue Agr. Exp. Sta. Bulletin 105, September.

_____. 1985. "Household Consumption of Storable Manufactured Dairy Products." Purdue Agr. Exp. Sta. Bulletin No. 85, June.

Brandow, G. E. 1961. "Interrelationships among Demand for Farm Products and Implications for Control of Market Supply." Pennsylvania Agr. Exp. Sta. Bulletin 680, August.

Dahlgran, Roger A. 1980. "Welfare Costs and Interregional Income Transfers Due to

Regulation of Dairy Markets." *American Journal of Agricultural Economics* 62(2):288–296.

George, P. S. and G. A. King. 1971. "Consumer Demand for Food Commodities in the United States with Projections for 1980." California Agr. Exp. Sta. Giannini Foundation Monograph No. 26, March.

Hallberg, M. C. 1982. "Cyclical Instability in the U.S. Dairy Industry without Government Regulations." *Agricultural Economic Research* 34(no.1):1–11.

Hallberg, M. C. and R. A. King. 1980. "Pricing Performance in the U.S. Dairy Industry." The Pennsylvania State University, Department of Agricultural Economics and Rural Sociology. A.E. & R.S. #150, June.

Hallberg, M. C. and R. F. Fallert. 1976. "Policy Simulation Model for the United States Dairy Industry." Pennsylvania Agr. Exp. Sta. Bulletin 805, January.

Heien, Dale M. 1982. "The Structure of Food Demand: Interrelatedness and Duality." American Journal of Agricultural Economics 64(2):213–221.

Heien, Dale M. and Cathy Roheim Wessells. 1988. "The Demand for Dairy Products: Structure, Prediction, and Decomposition." *American Journal of Agricultural Economics* 70(2):219–228.

Holt, C. N. 1965. "Quantitative Decision Analysis and National Policy: How Can We Bridge the Gap?" In *Quantitative Planning of Economic Policy*. ed. B. G. Hickman. Washington D.C.: The Brookings Institution.

Huang, Kuo S. 1985. "U.S. Demand for Food: A Complete System of Price and Income Effects." USDA/ERS Technical Bulletin No. 1714, December.

Kessel, Reuben. 1967. "Economic Effects of Federal Regulation of Milk Markets." *Journal of Law and Economics* 10(4):51–78.

Manchester, Alden C. 1983. *The Public Role in the Dairy Economy*. Boulder, Colo.: Westview.

Rojko, Anthony S. 1957. "The Demand and Price Structure for Dairy Products." USDA Technical Bulletin No. 1168.

Thraen, Cameron, Jerome Hammond, and Boyd M. Buxton.n.d. "An Analysis of Household Consumption of Dairy Products." Minnesota Agr. Exp. Sta. Bulletin No. 515.

Thraen, Cameron S. and Jerome W. Hammond. 1983. "Price Supports, Risk Aversion and U.S. Dairy Policy: An Alternative Perspective of the Long-Term Impacts." University of Minnesota Department of Agricultural and Applied Economics Economic Report ER83-9, June.

U.S. Department of Agriculture. *Dairy Products*. Various issues. Crop Reporting Board. Washington, D.C.: Statistical Reporting Service.

_____. *Milk Production, Disposition, and Income*. Various issues. Crop Reporting Board. Washington, D.C.: Statistical Reporting Service.

_____. *Report to Congress on the Dairy Promotion Program*. 1987. Washington, D.C.: U.S. Government Printing Office, July.

Warland, Rex H., Cathy Kassab, Robert O. Herrmann, and Blair J. Smith. 1989. "Who's Drinking More (or Less) Milk?" *Farm Economics*. Pennsylvania State University Cooperative Extension Service.

INDEX